Untrodden Peaks and Unfrequented Valleys. Second edition.

Amelia Blandford Edwards

The BiblioLife Network

This project was made possible in part by the BiblioLife Network (BLN), a project aimed at addressing some of the huge challenges facing book preservationists around the world. The BLN includes libraries, library networks, archives, subject matter experts, online communities and library service providers. We believe every book ever published should be available as a high-quality print reproduction; printed on- demand anywhere in the world. This insures the ongoing accessibility of the content and helps generate sustainable revenue for the libraries and organizations that work to preserve these important materials.

The following book is in the "public domain" and represents an authentic reproduction of the text as printed by the original publisher. While we have attempted to accurately maintain the integrity of the original work, there are sometimes problems with the original book or micro-film from which the books were digitized. This can result in minor errors in reproduction. Possible imperfections include missing and blurred pages, poor pictures, markings and other reproduction issues beyond our control. Because this work is culturally important, we have made it available as part of our commitment to protecting, preserving, and promoting the world's literature.

GUIDE TO FOLD-OUTS, MAPS and OVERSIZED IMAGES

In an online database, page images do not need to conform to the size restrictions found in a printed book. When converting these images back into a printed bound book, the page sizes are standardized in ways that maintain the detail of the original. For large images, such as fold-out maps, the original page image is split into two or more pages.

Guidelines used to determine the split of oversize pages:

• Some images are split vertically; large images require vertical and horizontal splits.
• For horizontal splits, the content is split left to right.
• For vertical splits, the content is split from top to bottom.
• For both vertical and horizontal splits, the image is processed from top left to bottom right.

UNTRODDEN PEAKS

AND

UNFREQUENTED VALLEYS

BY THE SAME AUTHOR.

UNIFORM WITH THIS VOLUME,

A THOUSAND MILES UP THE NILE.

With upwards of Seventy Illustrations, engraved on Wood by G. PEARSON,
after Finished Drawings executed on the spot by the Author.

A NEW EDITION, REVISED.

THE SASSO BIANCO FROM VAL. CORDEVOLE.

[*Frontispiece.*

UNTRODDEN PEAKS

AND

UNFREQUENTED VALLEYS

A Midsummer Ramble in the Dolomites

By AMELIA B. EDWARDS

AUTHOR OF "A THOUSAND MILES UP THE NILE," "LORD BRACKENBURY,"
"DEBENHAM'S VOW," "BARBARA'S HISTORY," &c., &c.

SECOND EDITION

PEASANT WOMAN OF LIVINALLUNGO

LONDON
GEORGE ROUTLEDGE AND SONS
BROADWAY, LUDGATE HILL
GLASGOW, MANCHESTER, AND NEW YORK
1890

LONDON :

BRADBURY, AGNEW, & CO., PRINTERS, WHITEFRIA

THE FIRST EDITION OF THIS BOOK

WAS DEDICATED

TO

MY AMERICAN FRIENDS IN ROME.

———

I DESIRE TO DEDICATE

THIS SECOND EDITION

TO

MY AMERICAN FRIENDS
IN ALL PARTS OF THE WORLD

PREFACE

TO THE SECOND EDITION.

———

IN preparing this Second Edition of a book which
has long been out of print, I have been careful to add
such information as may render it more useful to travellers in the Dolomite country. Some rough bridle-
paths have been superseded by good roads; some old
hostelries have been closed; some new inns have been
opened; and the approach to Cortina has been much
facilitated by the extension of the Conegliano line to
Vittorio, and by the network of new branch lines con-
necting Belluno, Feltre, and Bassano with the main
lines from Venice and Verona. Beyond these improve-
ments, little is changed since "L. and the Writer"
made their pleasant pilgrimage. The people are
almost as unsophisticated, and quite as friendly, as
ever; and if there should now be found less margin for

adventure of the mild kind described in the following pages, there is, by way of compensation, the certainty of better food and better accommodation than always fell to the lot of those who played the part of pioneers sixteen years ago.

I have thought it best to leave the original narrative unaltered, adding only a few foot-notes as to routes, inns, &c., where necessary. The new roads and lines of railway will be found duly entered in the map.

AMELIA B. EDWARDS.

WESTBURY-ON-TRYM,
May, 1889.

PREFACE

TO THE FIRST EDITION (1873).

———

THE district described in the following pages occupies that part of the South-Eastern Tyrol which lies between Botzen, Brunecken, Innichen, and Belluno. Within the space thus roughly indicated are found those remarkable limestone mountains called the Dolomites.

Till the publication of Ball's Guide to the Eastern Alps in 1868, and the appearance of Messrs. Gilbert and Churchill's joint volume in 1864,—the Dolomite district was scarcely known even by name to any but scientific travellers. A few geologists found their way now and then to Predazzo; a few artists, attracted in the first instance to Cadore as the birthplace of Titian, carried their sketch-books up the Ampezzo Thal; but there it ended. Even now, the general public is so slightly informed upon the subject that it is by no means uncommon to find educated persons who have never heard of the Dolomites at all, or who take them for a religious sect, like the Mormons or the Druses.

Nor is this surprising when we consider the nature of the ground lying within the area just named; the absence of roads; the impossibility of traversing the heart of the country, except on foot or on mule-back; the tedious postal arrangements; the want of telegraphic communication; and the primitive quality of the accommodation provided for travellers. A good road is the widest avenue to knowledge; but there is at present only one good and complete road in the whole district—namely, the strada regia which, traversing the whole length of the Ampezzo Thal, connects the Venetian provinces with Lower Austria. Other fragments of roads there are; but then they are only fragments, leading sometimes from point to point within an amphitheatre of mountains traversed only by mule-tracks.

When, however, one has said that there are few roads —that letters, having sometimes to be carried by walking postmen over a succession of passes, travel slowly and are delivered irregularly—that the inns are not only few and far between, but often of the humblest

kind—and that, except at Cortina, there is not a telegraph station in the whole country, one has said all that
can be said in disparagement of the district. For the
rest, it is difficult to speak of the people, of the climate,
of the scenery, without risk of being thought too partial
or too enthusiastic. To say that the arts of extortion
are here unknown—that the old patriarchal notion of
hospitality still survives, miraculously, in the minds of
the inn-keepers—that it is as natural to the natives of
these hills and valleys to be kind, and helpful, and
disinterested, as it is natural to the Swiss to be rapacious
—that here one escapes from hackneyed sights, from
overcrowded hotels, from the dreary routine of table
d'hôtes, from the flood of tourists,—is, after all, but to
say that life in the South-Eastern Tyrol is yet free from
all the discomforts which have of late years made
Switzerland unendurable; and that for those who love
sketching and botany, mountain-climbing and mountain
air, and who desire when they travel to leave London
and Paris behind them, the Dolomites offer a " playground " far more attractive than the Alps.

That a certain amount of activity and some power to
resist fatigue, are necessary to the proper enjoyment of
this new playground, must be conceded from the

beginning. The passes are too long and too fatiguing for ladies on foot, and should not be attempted by any who cannot endure eight and sometimes ten hours of mule-riding. The food and cooking, as will be seen in the course of the following narrative, are for the most part indifferent; and the albergos, as I have already said, are often of the humblest kind. The beds, however, in even "the worst inn's worst room" are generally irreproachable; and this alone covers a multitude of shortcomings. Anyone who has visited Ober-Ammergau during the performances of the Passion Play can form a tolerably exact idea of the sort of accommodation to be met with at Cortina, Caprile, Primiero, Predazzo, Paneveggio, Corfara, and St. Ulrich. A small store of tea, arrowroot, and Liebig's extract, a bottle or two of wine and brandy, a flask of spirits of wine and an Etna, are almost indispensable adjuncts to a lengthened tour in these mountains. The basket which contains them adds but little to the impedimenta, and immensely to the well-being of the traveller.

For ladies, side-saddles are absolutely necessary, there being only two in the whole country, and but one of these for hire. There is no need to take them out from England. They can always be bought at the

last large town through which travellers pass on their
way to the Dolomites, and sold again at the first they
come to on leaving the district.

Some knowledge of Italian and German is also indispensable. French here is of no use whatever; and
Italian is almost universally spoken. It is only in the
Grödner Thal, the Gader Thal, and the country north
of the Ampezzo, that one comes upon a purely German
population.

The Dolomite district is most easily approached from
either Venice, Botzen, or Brunecken. All that is
grandest, all that is most attractive to the artist, the
geologist, and the Alpine climber, lies midway between
these three points, and covers an area of about thirty-
five miles by fifty. The scenes which the present
writer has attempted to describe, all lie within that
narrow radius.

A word ought, perhaps, to be said with regard to the
title of this book, which, at first hearing, may be taken
to promise more than the author is prepared to fulfil.
But it means simply that here in South Tyrol, within
seventy-two hours of London, there may be found a
large number of yet " untrodden peaks," and a network
of valleys so literally "unfrequented" that we journeyed

sometimes for days together without meeting a single traveller either in the inns or on the roads, and encountered only three parties of English during the whole time between entering the country on the Conegliano side and leaving it at Botzen.

Of these unascended Dolomites, many exceed 10,000 feet in height; and some—as the Cima di Fradusta, the Pala di San Martino and the Sass Maor—are so difficult, that the mountaineer who shall first set foot upon their summits will have achieved a feat in no way second to that of the first ascent of the Matterhorn.

Of the nature and origin of Dolomite much has been written and much conjectured by French and German geologists; but nothing as yet seems definitely proved. The Coral Reef theory of Baron Richthofen seems, however, to be gaining general acceptance, and to the unscientific reader it sounds sufficiently conclusive. He grounds his theory upon certain facts, such as :—

1. The singular isolation of these mountains, many of which stand detached and alone, falling away steeply on all sides in a way that cannot be the result of any process of denudation.

2. The presence in their substance of such marine deposits as are found in the same position in the Coral

Reefs now in progress of formation in the Pacific and Indian Oceans, and on the Australian coast-line.

3. The absence of all deep-sea deposits.

4. The absence of all trace of volcanic origin.

5. The peculiarity of their forms, which reproduce in a remarkable manner the forms of the Coral Reef "Atolls" of the present day, being vertical, like huge walls, towards the wash of the tide, and supported on the lee side by sloping buttresses.

6. Their lines of curvature, and the kind of enclosures which they fence in; so again reproducing the construction of the Coral Reefs, which thus embay spaces of shallow water.

7. Finally, the multiform evidences (too numerous to be dwelt upon here) of how the Dolomite must have been slowly and steadily superimposed during long ages upon lower original beds of other rock, and the difficulty of accounting for this process by any other hypothesis.

"The Schlern," says Richthofen, taking this for his representative mountain, "is a Coral Reef; and the entire formation of Schlern Dolomite has in like manner originated through animal activity."*

* I am indebted to Mr. G. C. Churchill's admirable "Physical Description of the Dolomite District," for the particulars epitomized above.

The Dolomite derives its name from that of Monsieur Dolomieu, an eminent French savant of the last century, who travelled in South Tyrol somewhere about the years 1789 and 1790, and first directed the attention of the scientific world towards the structural peculiarities of this kind of limestone.

In conclusion, I can only add that I have tried to give a faithful impression of the country and the people; but that, having endeavoured when on the spot to sketch that which defied the pencil, so I fear that in the following pages I have striven to describe that which equally defies the pen.

AMELIA B. EDWARDS.

WESTBURY-ON-TRYM,
June 5, 1873.

CONTENTS.

———•———

CHAPTER I.

MONTE GENEROSO TO VENICE.

CHAPTER II.

VENICE TO LONGARONE.

b

CHAPTER III.

LONGARONE TO CORTINA.

CHAPTER IV.

AT CORTINA.

CHAPTER V.

CORTINA TO PIEVE DI CADORE.

CHAPTER VI.

AURONZO AND VAL BUONA.

CHAPTER VII.

CAPRILE.

CHAPTER XI.
THE FASSA THAL AND THE FEDAJA PASS.

CHAPTER XII.
THE SASSO BIANCO.

CHAPTER XIII.
FORNO DI ZOLDO AND ZOPPÉ.

CHAPTER XIV.

CAPRILE TO BOTZEN.

LIST OF ILLUSTRATIONS.

—•—

MAP.

FULL-PAGE ILLUSTRATIONS.

WOOD ENGRAVINGS IN THE TEXT.

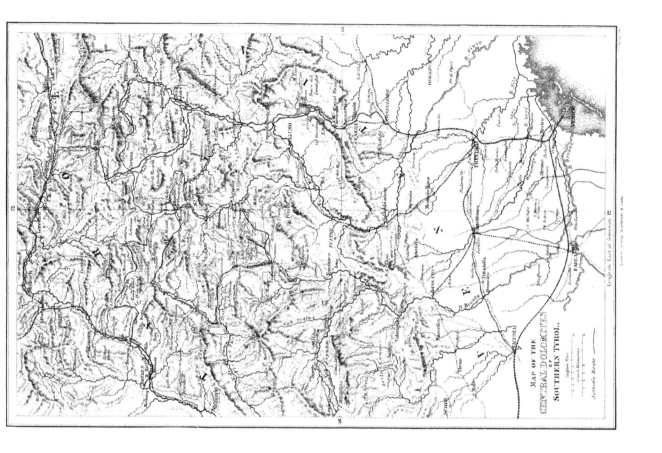

MAP OF THE
CENTRAL DOLOMITES
OR
SOUTHERN TYROL.

Author's Route

MONTE GENEROSO TO VENICE.

HOTEL AT MONTE GENEROSO—WONDERFUL PANORAMA — DREAMS
ABOUT DOLOMITES—DIFFICULTIES—THE REDOUTABLE COURIER—
THE REV. JOHN R.—CHOICE OF ROUTES—MENDRISIO TO COMO—
COMO CATHEDRAL—FELLOW TRAVELLERS ON BOARD THE STEAMER
— BELLAGGIO — LECCO TO BERGAMO, POST-HASTE — PANORAMIC
SCENERY BY RAIL—VENICE UNDER A NEW ASPECT—WE LAY IN
STORE OF PROVISIONS FOR THE COMING JOURNEY—THE QUESTION
OF SIDE-SADDLES—READY TO START.

UNTRODDEN PEAKS

AND

UNFREQUENTED VALLEYS.

CHAPTER I.

MONTE GENEROSO TO VENICE.

AN autumn in North Italy, a winter in Rome, a springtide in Sorrento, brought summer round again— the rich Italian summer, with its wealth of fruits and flowers, its intolerable heat, and its blinding brightness. The barbarian tide had long ago set northwards and overflowed into Switzerland. Even those who had lingered longest were fain at last to turn their faces towards the hills; and so it happened that the writer and a friend who had joined her of late in Naples, found themselves, about the middle of June, 1872, breathing the cooler airs of Monte Generoso.

Here was a pleasant hotel, filled to overflowing, and numbering among its guests many Roman friends of the past season. Here, too, were green slopes, and shady woods, and meadows splendid with such wild flowers as none of us had ever seen elsewhere. The steaming lakes, from which we had just escaped—Como, Lugano and Maggiore—lay in still, shining sheets three thousand feet below. The vast Lombard flats on the one side simmered all day in burning mists to the farthest horizon. The great snow-ranges bounding Switzerland and Tyrol on the other, glowed with the rose of every dawn, and turned purple when the sun went down behind them in glory every evening.

Having this wondrous panorama constantly before our eyes, with its changing lights and shadows, and its magical effects of cloud-wreath and shower—catching now a sudden glimpse of the Finsteraarhorn and the Bernese range—now an apparitional vision of Monte Rosa, or the Matterhorn, or even (on a clear morning, from the summit behind the hotel) of the far-distant Ortler Spitze on the Tyrolese border—we began, some-how, to think and talk less of our proposed tour in the Engadine; to look more and more longingly towards the north-eastern horizon; and to dream in a vague way of those mystic mountains beyond Verona which we knew of, somewhat indefinitely, as the Dolomites.

The Dolomites! It was full fifteen years since I had first seen sketches of them by a great artist not long since passed away, and their strange outlines and still stranger colouring had haunted me ever since. I thought of them as every summer came round; I regretted them every autumn; I cherished dim hopes about them every spring. Sketching about Venice in a gondola a year before the time of which I write, I used to be ever looking towards the faint blue peaks beyond Murano.

In short, it was an old longing; and now, high up on

the mountain side, with *Zermatt* and the Engadine close within reach, and the multitudinous Alps extending across half the horizon, it came back upon me in such force as to make all that these great mountains and passes had to show seem tame and undesirable.

Fortunately, my friend (whom I will call L. for brevity) had also read and dreamed of Dolomites, and was as eager to know more of them as myself; so we soon reached that stage in the history of every expedition when vague possibilities merge into planned certainties, and the study of maps and routes becomes the absorbing occupation of every day.

There were, of course, some difficulties to be overcome ; not only those difficulties of accommodation and transit which make the Dolomite district less accessible than many more distant places, but special difficulties arising out of our immediate surroundings. There was Sophia, for instance (L.'s maid), who, being delicate, was less able for mountain work than ourselves. And there was the supreme difficulty of the courier—a gentleman of refined and expensive tastes, who abhorred what is generally understood by " roughing it," despised primitive simplicity, and exacted that his employers should strictly limit their love of the picturesque to districts abundantly intersected by railways and well furnished with first-class hotels.

That this illustrious man should look with favour on our new project was obviously hopeless ; so we discussed it secretly " with bated breath," and the proceedings at once assumed the delightful character of a conspiracy. The Rev. John R., who had been acting for some weeks as English chaplain at Stresa, was in the plot from the beginning.   He had himself walked through part of our Dolomite route a few years before, and so gave us just that sort of practical advice which is, of all help in travelling, the most valuable.   For this; for his gallant indifference to the ultimate wrath of the courier; and for the energetic way in which (with a noble disregard of appearances, for which we can never be sufficiently grateful) he made appointments with us in secluded summerhouses, and attended stealthy indoor conferences at hours when the servants were supposed to be at meals, I here beg to offer him our sincere and hearty thanks.

All being at last fully planned, it became necessary to announce our change of route.   The great man was accordingly summoned; the writer, never famous for moral courage, ignominiously retreated; and L., the dauntless, undertook the service of danger.   Of that tremendous interview no details ever transpired.   Enough that L. came out from it composed but victorious; and that the great man, greater than ever under defeat, comported himself thenceforth with such a nicely adjusted air of martyrdom and dignity as defies description.

Now, there are three ways by which to enter the Dolomite district; namely, by Botzen, by Brunecken, or by Venice; and it fell in better with our after plans to begin from Venice.   So on Thursday the 27th of June, we bade farewell to our friends on Monte Generoso, and went down in all the freshness and beauty of the early morning.   It was a day that promised well for the beginning of such a journey. There had been a heavy thunderstorm the night before, and the last cumuli were yet rolling off in a long billowy rack upon the verge of land and sky. The plains of Lombardy glittered wide and far; Milan gleamed, a marble-speck, in the mid-distance; and farthest seen of all, a faint, pure obelisk of snow, traced as it were upon the transparent air, rose Monte Viso, a hundred and twenty miles away.

But soon the rapidly descending road and thickening woods shut out the view, and in less than two hours we were down again in Mendrisio, a clean little town containing an excellent hotel, where travellers bound for the mountain, and travellers coming down to the plains, are wont to rest. Here we parted from our heavy luggage, keeping only a few small bags for use during the tour. Here also we engaged a carriage to take us on to Como, where we arrived about midday, after a dull and dusty drive of some two hours more.

It was our intention to push on that afternoon as far as Bellaggio, and in the morning to take the early steamer to Lecco, where we hoped to catch the 9.25 train reaching Venice at 4.30. Tired as we now were, it was pleasant to learn that the steamer would not leave till three, and that we might put up for a couple of hours at the Hotel Volta—not only the best in Como, but one of the best in Italy. Here we rested and took luncheon, and, despite the noontide blaze out of doors, contrived to get as far as that exquisite little miniature in marble, the Cathedral. Lingering there till the last moment, examining the cameo-like bas-reliefs of the façade, the strange beasts of unknown date that support the holy-water basins near the entrance, and the delicate

Italian-Gothic of the nave and aisles, we ran back just in time to see our effects being wheeled down to the pier, and to find the steamer not only crowded with passengers, but the deck piled, funnel-high, with bales of raw silk, empty baskets, and market produce of every description.

We were the only English on board, as we had been the only English in the streets, in the hotel, and apparently in all the town of Como. Our fellow-passengers were of the bourgeois class—stout matrons with fat brown hands cased in netted mittens and loaded with rings; elderly *pères de famille* in straw hats; black-eyed young women in gay shawls and fawn-coloured kid boots; and a sprinkling of priests. It had probably been market-day in Como; for the fore-deck was crowded with chattering country folk, chiefly bronzed women in wooden clogs, some few of whom wore in their plaited hair that fan-shaped head-dress of silver pins, which, though chiefly characteristic of the Canton Tessin, just over the neighbouring Swiss border, is yet worn all about the neighbourhood of the lakes.

So the boat steamed out of the little port and along the glassy lake, landing many passengers at every stage; and the fat matrons drank iced Chiavenna beer; and the priests talked together in a little knot, and made merry among themselves. There were three of them—one rubicund, jovial, and somewhat threadbare; another very bent, and toothless, and humble, and desperately shabby; while the third, in shining broadcloth and a black satin waistcoat, carried himself like a gentleman and a man of the world, was liberal with the contents of his silver snuff-box, and had only to open his lips to evoke obsequious laughter.   We landed the two first at small water-side hamlets by the way, and the last went ashore at Cadenabbia, in a smart boat with two rowers.

Wooded hills, vineyards, villages, terraced gardens, gleaming villas bowered in orange groves, glided past meanwhile—a swift and beautiful panorama.   The little voyage was soon over, and the sun was still high when we reached Bellaggio ; a haven of delicious rest, if only for a few hours.

Next morning, however, by a quarter past seven, we were again on board and making, too slowly, for Lecco, where we arrived just in time to hear the parting whistle of the 9.25 train.   Now, as there were only two departures a day from this place and the next train would not start for seven hours, arriving in Venice close upon eleven at night, our case looked serious.   We drove, however, to an hotel, apparently the best ; and here the landlady, a bright energetic body, proposed that we should take a carriage across the country to Bergamo, and there catch up the 11.13 express from Milan.   Here was the carriage standing ready in the courtyard ; here were the horses ready in the stables ; here was her nephew ready to drive us—the lightest carriage, the best horses, the steadiest whip in Lecco !

Never was there so brisk a little landlady.   She allowed us no time for deliberation ; she helped to put the horses in with her own hands ; and she packed us off as eagerly as if the prosperity of her hotel depended on getting rid of her customers as quickly as possible.   So away we went, counting the kilometres against time all the way, and triumphantly rattling up to Bergamo station just twenty minutes before the express was due.

Then came that well-known route, so full of beauty, so rich in old romance, that the mere names of the stations along the line make Bradshaw read like a page of poetry—Brescia, Verona, Vicenza, Padua, Venice. For the traveller who has gone over all this ground at his leisure and is familiar with each place of interest as it flits by, I know no greater enjoyment than to pass them thus in rapid review, taking the journey straight through from Milan to Venice on a brilliant summer's day. What a series of impressions! What a chain of memories! What a long bright vision of ancient cities with forked battlements; white convents perched on cypress-planted hills; clustered villages, each with its slender campanile; rock-built citadels, and crumbling mediæval towns; bright rivers, and olive woods, and vineyards without end; and beyond all these a background of blue mountains ever varying in outline, ever changing in hue, as the clouds sail over them and the train flies on!

By five o'clock we were in Venice. I had not thought, when I turned southwards last autumn, that I should find myself threading its familiar water-ways so soon again. I could hardly believe that here was the Grand Canal, and yonder the Rialto, and that those white domes now coming into sight were the domes of Santa Maria della Salute. It all seemed like a dream.

And yet, somehow, it was less like a dream than a changed reality. It was Venice; but not quite the old Venice. It was a gayer, fuller, noisier Venice; a

Venice empty of English and American tourists; full to overflowing of Italians in every variety of summer finery; crowded with artists of all nations sketching in boats, or surrounded by gaping crowds in shady corners and porticos; a Venice whose flashing waters were now cloven by thousands of light skiffs with smart striped awnings of many colours, but whence the hearse-like, tufted gondola, so full of mystery and poetry, had altogether vanished; a Venice whose every side-canal swarmed with little boys learning to dive, and with swimmers of all ages; where dozens of cheap steamers (compared with which the Hungerford penny boats would seem like floating palaces) were hurrying to and fro every quarter of an hour between the Riva dei Schiavone and the bathing-places on the Lido; a Venice in which every other house in every piazza had suddenly become a café; in which brass bands were playing, and *caramels* were being hawked, and iced drinks were continually being consumed from seven in the morning till any number of hours after midnight; a Venice, in short, which was sunning itself in the brief gaiety and prosperity of the bathing season, when all Italy north of the Tiber, and a large percentage of strangers from Vienna, St. Petersburg, and the shores of the Baltic, throng thither to breathe the soft sea-breezes off the Adriatic.

We stayed three days at Danieli's, including Sunday; and, mindful that we were this time bound for a district where roads were few, villages far between, and inns scantily provided with the commonest necessaries, we took care to lay in good store of portable provision for the journey. Our Saturday and Monday were therefore spent chiefly in the mazes of the Merceria. Here we bought two convenient wicker-baskets, and wherewithal to stock them—tea, sugar, Reading biscuits in tins, chocolate in tablets, Liebig's Ramornie extract, two bottles of Cognac, four of Marsala, pepper, salt, arrow-root, a large metal flask of spirits of wine, and an Etna. Thus armed, we could at all events rely in case of need upon our own resources; and of milk, eggs, and bread we thought we might make certain everywhere. Time proved, however, that in the indulgence of even this modest hope we over-estimated the fatness of the land; for it repeatedly happened that (the cows being gone to the upper pastures) we could get no milk; and on one memorable occasion, in a hamlet containing at least three or four hundred souls, we could get no bread.

There was yet another point upon which we were severely "exercised," and that was the question of side-saddles. Mr. R., on Monte Generoso, had advised us to purchase them and take them with us, doubting whether we should find any between Cortina and Botzen. Another friend, however, had positively assured us of the existence of one at Caprile; and where there was one, we hoped there might be two more. Anyhow, we were unwilling to add the bulk and burden of three side-saddles to our luggage; so we decided to go on, and take our chance. I suspect, however, that we had no alternative, and that one might as well look for skates in Calcutta as for saddlery in Venice. As the event proved, we did ultimately succeed in capturing two side-saddles (the only two in the whole district), and in forcibly keeping them throughout the journey; but this was a triumph of audacity, never to be repeated. Another time, we should undoubtedly provide ourselves with side-saddles either at Padua or Vicenza on the one side, or at Botzen on the other.

By Monday evening the 1st of July, our preparations were completed; our provision baskets packed; our stores of sketching and writing materials duly laid in; and all was at length in readiness for an early start next morning.

VENICE TO LONGARONE.

CHAPTER II.

VENICE TO LONGARONE.

HAVING risen at grey dawn, breakfasted at a little after 5 A.M., and pulled down to the station before half the world of Venice was awake, it was certainly trying to find that we had missed our train by about five minutes, and must wait four hours for the next. Nor was it much consolation, though perhaps some little relief, to upbraid the courier who had slept too late, and so caused our misfortune. Sulky and silent, he piled our bags in a corner and kept gloomily aloof; while we, cold, dreary, and discontented, sat shivering in a draughty passage close against the ticket office, counting the weary hours and excluded even from the waiting-rooms, which were locked up " per ordine superiore " till half an hour before the time at which we now could proceed upon our journey. The time, however, dragged by somehow, and when at ten o'clock we at last found ourselves moving slowly out of the station, it seemed already like the middle of the day.

And now again we traversed the great bridge and the long, still, glassy space of calm lagune, and left the lessening domes of Venice far behind. And now, Mestre station being passed and the firm earth reached again, we entered on a vast flat all green with blossoming Indian corn, and intersected by a network of broad dykes populous with frogs. Heavens! how they croaked! Driving out from Ravenna to Dante's famous pine-forest the other day, we had been almost deafened by them; but the shrill chorus of those Ravenna frogs was as soft music compared with the unbridled revelry of their Venetian brethren. These drowned the very noise of the train, and reduced us to dumb show till we were out of their neighbourhood.

So we sped on, the grey-blue mountains that we had been looking at so longingly from Venice these last three days growing gradually nearer and more definite. Soon we begin to distinguish a foreground of lower hill-tops, some dark with woods, others cultivated from base to brow and dotted over with white villages. Then by-and-by comes a point, midway as it were between Venetia and Tyrol, whence we see the last tapering Venetian campanile outlined against the horizon on the one hand, and the first bulbous Tyrolean steeple, shaped like the morion of a mediæval man-at-arms, peeping above the roof of a little hill-side hamlet on the other.

The dykes and frogs are now left far behind; the line is bordered on both sides by feathery acacia hedges; and above the lower ranges of frontier mountains, certain strange jagged peaks, which, however, are not Dolomite, begin to disengage themselves from the cloudy background of the northern sky. No, they cannot be Dolomite, though they look so like it; for we have been told

that we shall see no true Dolomite before to-morrow.
It is possible, however, as we know, to see the Antelao
from Venice on such a clear day as befalls about a
dozen times in the course of a summer; but here, even
if the sky were cloudless, we are too close under the
lower spurs of the outlying hills to command a view of
greater heights beyond.

Treviso comes next—apparently a considerable place.
Here, according to Murray, is a fine Annunciation of
Titian to be seen in the Duomo, but we, alas! have no
time to stay for it. Here also, as our fellow-traveller,
the priest in the corner, says unctuously, opening his
lips for the first and last time during the journey,“ they
make good wine.” (“ Qui si fa buon vino.”)

At Treviso we drop a few third-class travellers, and
(being now just eighteen miles from Venice, and exactly
half-way to Conegliano) go on again through a fat, flat
country, past endless fields of maize and flax; past
trailing vines reared, as in the Tyrol, on low slanting
trellises close against the ground; past rich midsummer
meadows where sunburnt peasants wade knee-deep in
wild-flowers, and their flocks of turkeys are guessed at
rather than seen; past villages, and small stations, and
rambling farmhouses, and on towards the hills that are
our goal. By-and-by, some four or five miles before
Conegliano, the fertile plain is scarred by a broad tract
of stones and sand, in the midst of which the Piave,
grey, shallow, and turbid, hurries towards the sea. Of
this river we are destined to see and know more hereafter, among its native Dolomites.

And now we are at Conegliano, the last point to

which the railway can take us,* and which, in consequence of our four hours' delay this morning, we have
now no time to see. And this is disappointing; for
Conegliano must undoubtedly be worth a visit. We
know of old Palazzos decorated with fast-fading frescoes
by Pordenone; of a theatre built by Segusini; of an
altar-piece in the Duomo by Cima of Conegliano, an
exquisite early painter of this place, whose works are
best represented in the Brera of Milan, and whose
clear, dry, polished style holds somewhat of an intermediate place between that of Giovanni Bellini and
Luca Signorelli.

But if we would reach Longarone—our first stopping
place—to-night, we must go on; so all we carry away
is the passing remembrance of a neat little station; a
bright, modern-looking town about half a mile distant;
a sprinkling of white villas dotted over the neighbouring
hill-sides; and a fine old castle glowering down from a
warlike height beyond.

And now the guard's whistle shrills in our ears for the
last time for many weeks, and the train, bound for
Trieste, puffs out of the station, disappears round a
curve, and leaves us on the platform with our pile of
bags at our feet and all our adventures before us. We
look in each other's faces. We feel for the moment as

---

* There is now a branch line from Conegliano to Vittorio (see *p.* 47),
which considerably shortens the journey by road for those who desire to go
by way of the Lago di S. Croce. There is, however, a yet quicker route
from Venice by the new line from Treviso to Belluno, which makes it
possible for the traveller to reach Cortina in one long day. From Belluno,
a diligence runs twice a day to Longarone, whence a carriage can be taken
to Cortina. (*Note to Second Edition.*)

Martin Chuzzlewit may have felt when the steamer landed him at Eden, and there left him. Nothing, in truth, can be more indefinite than our prospects, more vague than our plans. We have Mayr's maps, Ball's Guide to the Eastern Alps, Gilbert and Churchill's book, and all sorts of means and appliances; but we have not the slightest idea of where we are going, or of what we shall do when we get there.

There is, however, no time now for misgivings, and in a few minutes we are again under way. Some three or four dirty post-omnibuses and bilious-looking yellow diligences are waiting outside, bound for Belluno and Longarone; also one tolerable carriage with a pair of stout grey horses, which, after some bargaining, is engaged at the cost of a hundred lire.* For this sum the driver is to take us to-day to Longarone, and to-morrow to Cortina in the Ampezzo Valley—a distance, altogether, of something like seventy English miles. So the bags are stowed away, some inside, some outside; and presently, without entering the town at all, we drive through a dusty suburb and out again upon the open plain.

A straighter road across a flatter country it would be difficult to conceive. Bordered on each side by a row of thin poplars, and by interminable fields of Indian corn, it goes on for miles and miles, diminishing to a point in the far distance, like the well-known diagram of an avenue in perspective. And it is the peculiar attribute of this point to recede steadily in advance of us, so that we are always going on, as in a dreadful

* About four pounds English.

dream, and never getting any nearer. As for incidents by the way, there are none. We pass one of the lumbering yellow diligences that were standing erewhile at Conegliano station; we see a few brown women hoeing in the Indian corn; and then for miles we neither pass a house nor meet a human being.

It appears to me that hours must have gone by thus when I suddenly wake up, baked by the sun and choked by the dust, to find the whole party asleep, driver included, and the long distant hills now rising close before us. Seeing a little town not a quarter of a mile ahead—a little town bright in sunshine against a background of dark woods, with a ruined castle on a height near by, I know at once that this must be Ceneda—the Ceneda that Titian loved—and that yonder woods and hills and ruined castle are the same he took for the landscape background to his St. Peter Martyr. Here he is said to have owned property in land; and at Manza, four miles off, he built himself a summer villa.

Now, moved by some mysterious instinct, the driver wakes up just in time to crack his whip, put his horses into a gallop, and clatter, as foreign vetturini love to clatter, through the one street which is the town. But in vain; for Ceneda—silent, solitary, basking in the sun, with every shutter closed and only a lean dog or two loitering aimlessly about the open space in front of the church—is apparently as sound asleep as an enchanted town in a fairy tale. Not a curtain is put aside, not a face peers out upon us as we rattle past. The very magpie in his wicker cage outside the barber's shop is dozing on his perch, and scarcely opens an eye, though we make noise enough to rouse the Seven Sleepers.

Once past the houses, we fall back, of course, into the old pace, the gracious hills drawing nearer and unfolding fresh details at every step. And now, at last, green slopes and purple crags close round our path ; the road begins to rise ; a steep and narrow gorge, apparently a mere cleft in the mountains like the gorge of Pfeffers, opens suddenly before us ; and from the midst of a nest of vines, mulberry trees and chestnuts, the brown roofs and campaniles of Serravalle lift themselves into sight.

Serravalle, though it figures on the map in smaller type than Ceneda, which is, or was, an Episcopal residence, is yet a much more considerable place, covering several acres, and straggling up into the mouth of the gorge through which the Meschio comes hurrying to the plain. Strictly speaking, perhaps, there is now no Ceneda and no Serravalle, the two townships having been united of late by the Italian Government under the name of Vittorio ; but they lie a full mile apart, and no one seems as yet to take kindly to the new order of things.

Again our driver cracks his whip and urges his horses to a canter; and so, with due magnificence, we clatter into the town—a quaint, picturesque, crumbling, world-forgotten place, with old stone houses abutting on the torrent; and a Duomo that looks as if it had been left unfinished three hundred years ago ; and gloomy arcades vaulting the footways on each side of the principal street, as in

Strasburg and Berne. Dashing across the bridge and into the Piazza, we pull up before one of the two inns which there compete for possession of the infrequent traveller; for Serravalle boasts not only a Piazza and a Duomo, but two alberghi, two shabby little cafés, a Regia Posta, and even a lottery office with "Qui si giuoca per Venezia" painted in red letters across the window.

Here, too, the inhabitants are awake and stirring. They play at dominos in their shirt-sleeves outside the cafés. They play at "morra" in the shade of doorways and arcades. They fill water-jars, wash lettuces, and gossip at the fountain. They even patronise the drama, as may be seen by the erection of a temporary puppet-theatre ("patronised by His Majesty the King of Italy and all the Sovereigns of Europe") on a slope of waste ground close against the church. Nor is wanting the usual score or two of idle men and boys who immediately start up from nowhere in particular, and swarm, open-mouthed, about the carriage, staring at its occupants as if they were members of a travelling menagerie.

But Serravalle has something better than puppets and an idle population to show. The Duomo contains a large painting of the Madonna and Child in glory, by Titian, executed to order some time between the years 1542 and 1547—a grand picture belonging to what may perhaps be called the second order of the master's greatest period, and of which it has lately been said by an eminent traveller and critic that "it would alone repay a visit to Serravalle, even from Venice." With respect to the treatment of this fine work, Mr. Gilbert, whose admirable book on Titian and Cadore leaves nothing for any subsequent writer to add on these subjects, says :—" It is one of the grandest specimens of the master, and in very fair preservation. It represents the Virgin and Child in glory surrounded by angels, who fade into the golden haze above. Heavy-volumed clouds support and separate from earth this celestial vision ; and below, standing on each side, are the colossal and majestic figures of St. Andrew and St. Peter ; the former supporting a massive cross, the latter holding aloft, as if challenging denial of his faithfulness, the awful keys. Between these two noble figures, under a low horizon line, is a dark lake amidst darker hills, where a distant sail recalls the fisherman and his craft. Composition, drawing, colour, are all dignified and worthy of the master." *Cadore*, p. 43.

And now, time pressing, the day advancing, and three-fourths of the drive yet lying before us, we must push on, or Longarone will not be reached ere nightfall. So, having been sufficiently stared at—not only by the population generally, but by the landlord and landlady and everybody connected with the inn, as well as by the domino players, who leave their games to take part in the entertainment—we clatter off again and make straight for the rocky mouth of the gorge, now closing in upon, and apparently swallowing up, the long line of old stone houses creeping into the defile. Some of these, shattered and decaying as they are, show traces of Venetian-Gothic in pointed ogive window and delicate twisted column. They belonged, no doubt, to wealthy owners in the days when Titian used to ride over from Manza to visit his married daughter who lived at Serravalle.

Where the houses end, the precipices so close in that there is but just space for the road and the torrent. Then the gorge gradually widens through wooded slopes and hanging chestnut groves; farmhouses and châlets perched high on grassy plateaux begin to look more Swiss than Italian; mountains and forests all round shut in the view; and about two miles from Serravalle, the Meschio expands into a tiny, green, transparent lake, tranquil as a cloudless evening sky, and fringed by a broad border of young flax. A single skiff, reflected upside down as in a mirror, floats idly in the middle of the lake. The fisherman in it seems to be asleep. Not a ripple, not a breath, disturbs the placid picture in the water. Every hill and tree is there, reversed; and every reed is doubled.

This delicious pool, generally omitted in the maps, is the Lago di Serravalle. Woods slope down to the brink on one side, and the road, skirting the débris of an old landslip, winds round the other. Two tiny white houses with green jalousies and open Italian balconies at the head of the lake, a toy church on a grassy knoll, and a square mediæval watchtower clinging to a ridge of rock above, make up the details of a picture so serene and perfect that even Turner at his sunniest period could scarcely have idealized it.

The gorge now goes on widening and becomes a valley, once the scene of a bergfall so gigantic that it is

supposed to have turned the course of the Piave
(flowing out till then by Serravalle) and to have sent it
thenceforward and for ever through the Val di Mel.
This catastrophe happened ages ago—most probably in
pre-historic times ; yet the great barrier, six hundred
feet in height from this side, looks as if it might be less
than a century old. Few shrubs have taken root in
these vast hillocks of slaty débris, among and over
which the road rises continually. Few mosses have
gathered in the crannies of these monster blocks, which
lie piled like fallen towers by the wayside. All is bare,
ghastly, desolate.

As we mount higher, the outlying trees of a great
beech-forest on the verge of a lofty plateau to the right,
are pointed out by the driver as the famous Bosco del
Consiglio—a name that dates back to old Venetian
rule, when these woods furnished timber to the state.
Hence came the wood of which the " Bucentaur " was
built ; and—who knows ?—perhaps the merchant ships
of Antonio, and the war-galley in which " blind old
Dandolo " put forth against the Turk.

Presently, being now about four miles from
Serravalle, and the top of the great bergfall not yet
reached, we come upon another little green, clear lake,
about the size of the last—the Lago Morto. It lies
down in a hollow below the road, close under a huge,
sheer precipice blinding white in the sunshine, whence
half the mountain side looks as if it had been sliced
away at a blow. If it were not that the débris could
hardly be piled up where and how it is, leaving that
hollow in which the lake lies sleeping, one would

suppose this to be the spot whence the rock-slip came that time it barred out the Piave from the gorge of Serravalle.

According to the local legend, no boat can live upon those tranquil waters, and no bather who plunges into them may ever swim back to shore. Both are, in some terrible way, drawn down and engulphed " deeper than did ever plummet sound." It is said, however, that the last Austrian Governor of Lombardo-Venetia, being anxious to put an end to this superstition, brought up a boat from the Santa Croce side, and, in the presence of a breathless crowd from all the neighbouring villages, himself rowed the pretty wife of the Fadalto postmaster across the lake, and landed her triumphantly upon the opposite shore. Your Tyrolean peasant, however, is not easily disabused of ancient errors, and the Lago Morto, I am told, notwithstanding that public rehabilitation, enjoys its evil reputation to this day.

At length, having the Bosco del Consiglio always to the right, and the Col Vicentino with its scattered snow drifts towering to the left, we gain the summit of the ridge and see the lake of Santa Croce, looking wonderfully like the lake of Albano, lying close beneath our feet. Great mountains, all grey and purple crags above, all green corn-fields and wooded slopes below, enclose it in a nest of verdure. The village and church of Santa Croce, perched on a little grassy bluff, almost overhang the water. Other villages and campaniles sparkle far off on shore and hillside ; while yonder, through a gap in the mountains at the farther end of the lake, we are startled by a strange apparition of pale fantastic peaks lifted high against the northern horizon.

"Ecco!" says the driver, pointing towards them with his whip, and half turning round to watch the effect of his words, "Ecco i nostri Dolomiti!"

LAKE OF SANTA CROCE.

The announcement is so unexpected, that for the first moment it almost takes one's breath away. Having been positively told that no Dolomites would come into sight before the second day's journey, we have neither been looking for them nor expecting them—and yet there they are, so unfamiliar, and yet so unmistakeable! One feels immediately that they are unlike all other mountains, and yet that they are exactly what one expected them to be.

"Che Dolòmiti sono? Come si chiamano?" (What Dolomites are they? What are their names?) are the eager questions that follow.

But the bare geological fact is all our driver has to tell. They are Dolomites—Dolomites on the Italian side of the frontier. He knows no more; so we can only turn to our maps, and guess, by comparison of distances and positions, that those clustered *aiguilles* belong most probably to the range of Monte Sfornioi.

At Santa Croce we halt for half an hour before the door of an extremely dirty little albergo, across the front of which is painted in conspicuous letters, "Qui si vende buon vino a chi vuole."

Leaving the driver and courier to test the truth of this legend, we order coffee and drink it in the open air. The horses are taken out and fed. The writer, grievously tormented by a plague of flies, makes a sketch under circumstances of untold difficulty, being presently surrounded by the whole population of the place, among whom are some three or four handsome young women with gay red and yellow handkerchiefs bound round their heads like turbans. These damsels are by no means shy. They crowd; they push; they chatter; they giggle. One invites me to take her portrait. Another wishes to know if I am married. A third discovers that I am like a certain Maria Rosa whom they all seem to know; whereupon every feature of my face is discussed separately, and for the most part to my disparagement.

At this trying juncture, L., in a moment of happy inspiration, offers to show them the chromo-lithographs in Gilbert and Churchill's book, and so creates a diversion in my favour. Meanwhile the flies settle upon me in clouds, walk over my sky, drown themselves in the water bottles, and leave their legs in the brown madder; despite all which impediments, however, I achieve my sketch, and by the time the horses are put to, am ready to go on again.

The road now skirts the lake of Santa Croce, at the head of which extends an emerald-green flat wooded with light, feathery, yellowish poplars—evidently at one time part of the bed of the lake, from which the waters have long since retreated. From this point, we follow the line of the valley, passing the smart new village of Cadola; and at Capo di Ponte, whence the valley of Serravalle and the Val di Mel diverge at right angles, come again upon the Piave, now winding in and out among stony hillocks, like the Rhone at Leuk, and milk-white from its glacier-source in the upper Dolomites. The old bridge at Capo di Ponte—the old bridge which dated from Venetian times—is now gone; and with it the buttresses adorned with the lion of St. Mark mentioned by Ball and alluded to in Mr. Gilbert's " Cadore." Fragments of the ancient piers may yet be traced; but a new and very slight-looking iron bridge now spans the stream some fifty yards higher up. At Capo di Ponte, the most unscientific observer cannot fail to see that the Piave must once upon a time (most probably when the great bergfall drove its waters back from Serravalle) have here formed another lake, the great natural basin of which yet remains, with the river flowing through it in a low secondary channel.

And now the road enters another straight and narrow valley—the valley of the Piave—closed in far ahead by a rugged Dolomite, all teeth and needle-points. By this time the long day is drawing to a close. Cows after milking are being driven back to pasture ; labourers are plodding homewards ; and a party of country girls with red handkerchiefs upon their heads, wading knee-deep through the wild-flowers of a wayside meadow, look like a procession of animated poppies. Then the sun goes down ; the sky and the mountains turn cold and grey ; and just before the dusk sets in, we arrive at Longarone.

A large rambling village with a showy renaissance church and a few shabby shops—a big desolate inn with stone staircases and stone floors—a sullen landlord—a frightened, bare-footed chambermaid who looks as if she had just been caught wild in the mountains—bedrooms like barns, floors without carpets, windows without curtains—such are our first comfortless impressions of Longarone. Nor are these impressions in any wise modified by more intimate acquaintance. We dine in a desert of sitting-room at an oasis of table, lighted by a single tallow candle. The food is indifferent and indifferently cooked. The wine is the worst we have had in Italy.

Meanwhile, a stern and ominous look of satisfaction settles on the countenance of the great man whom we have so ruthlessly torn from the sphere which he habitually adorns. " I told you so " is written in every line of his face, and in the very bristle of his moustache. At last, being dismissed for the night and told at what hour to have the carriage round in the morning, he can keep silence no longer.

"We shall not meet with many inns so good as this, where we are going," he says, grimly triumphant. "Good night, ladies!"—and with this parting shot, retires.

My bedroom that night measures about thirty-five feet in length by twenty-five in breadth, and is enlivened by five windows and four doors. The windows look out variously upon street, courtyard, and stables. The doors lead to endless suites of empty, shut-up rooms, and all sorts of intricate passages. 'Tis as ghostly, echoing, suicidal a place to sleep in as ever I saw in my life!

LONGARONE TO CORTINA.

THE PIC GALLINA—A COMMUNICATIVE PRIEST—THE TIMBER TRADE—
THE SMALLEST CHURCH IN ITALY—CASTEL LAVAZZO—PERAROLO—A
VISION OF THE ANTELAO—THE ZIGZAG OF MONTE ZUCCO—TAI CADORE
—ONE OF THE FINEST DRIVES IN EUROPE—THE GLORIES OF THE
AMPEZZO THAL—THE PELMO—THE ROCHETTA—THE LANDSLIP OF
1816—THE ANTELAO—THE CRODA MALCORA—SORAPIS—WE CROSS
THE AUSTRIAN FRONTIER—THE BEC DI MEZZODI—THE TOFANA—
MONTE CRISTALLO—CORTINA—ARRIVAL AT GHEDINA'S INN—"IL
TUCKETT'S" NAME PROVES A WORD OF MIGHT—A THOROUGH TYRO-
LEAN HOSTELRY—PREPARATIONS FOR THE *SAGRO*.

CHAPTER III.

LONGARONE TO CORTINA.

LONGARONE, seen at six o'clock on a grey, dull morning, looked no more attractive than at dusk the evening before. There had been thunder and heavy rain in the night, and now the road and footways were full of muddy pools. The writer, however, was up betimes, wandering alone through the wet streets; peeping into the tawdry churches; spelling over the framed and glazed announcements of births, deaths, and marriages at the Prefettura; sketching the Pic Gallina, a solitary conspicuous peak over against the mouth of the Val Vajont, on the opposite bank of the Piave; and seeking such scattered crumbs of information as might fall in her way.

To sketch, even so early as six A.M., without becoming the nucleus of a crowd, is, of course, impossible; and the crowd this time consisted of school children of all ages, quite as " untameable," and almost as numerous, as the flies of Santa Croce. Presently, however, came by a mild, plump priest in a rusty *soutane*, who chased the truants off to the parish school-house, and himself lingered for a little secular chat by the way.

He had not much to tell ; yet he told the little that
he knew pleasantly and readily. The parish, he said,
numbered about three thousand souls—a pious, industrious folk mainly supported by the timber trade, which

PIC GALLINA.

is the staple of these parts. This timber, being cut,
sold, and branded in the Ampezzo Thal, is floated
down the Boita to its point of confluence with the
Piave at Perarolo, and thence, carried by the double
current, comes along the valley of the Piave and the

Val di Mel, to be claimed by its several purchasers along the banks, and caught as it passes by. Thus it is that every village by the way is skirted by saw-mills and timber-yards, and that almost every man is a carpenter. He then went on to tell me that my peak was called the Pic Gallina, or Hen's beak; that there existed a practicable short cut for pedestrians by way of the Val Vajont to Udine and the Trieste railway; that the " gran' Tiziano " was born on the banks of the Piave higher up, at Pieve di Cadore; that the Dolomites were the highest mountains in the world (which I am afraid I pretended to believe); that the large church in the Piazza was the church of the Concezione; that the little church at the back, dedicated to San Liberale, was the smallest church in Italy (which no doubt was true, seeing that you might put it inside St. Lawrence, Undercliff, and yet leave a passage to walk round); and finally, that Castel Lavazzo, seen from a point about a quarter of a mile farther on, was the most picturesque view in the valley, and the best worth sketching. Having delivered himself of which information, apocryphal and otherwise, he lifted his shovel-hat with quite the air of a man of the world, and bade me good morning.

Of course I went at once in search of the view of Castel Lavazzo, and finding it really characteristic of the Val di Piave, succeeded in sketching it before it was time to return to breakfast.

By nine, we were on the road again, following the narrow gorge which was soon to lead us into the real world of Dolomite. The morning was now alternately bright and showery, and the dark, jagged peaks that closed in the distance were of just that rich, deep, incredible ultra-marine blue that Titian loved and painted so often in his landscape backgrounds.

CASTEL LAVAZZO.

At Termine, a little timber-working hamlet noisy with saw-mills, about a mile beyond Castel Lavazzo, the defile narrows so suddenly that one gigantic grey and golden crag seems to block the end of the village street. The women here are handsome, and wear folded cloths upon their heads as in the hills near Rome; and the men wear wooden clogs, as at Lugano. A slender waterfall wavers down the face of a cliff on the opposite side of the river. Primitive breakwaters, like huge baskets of rude wicker-work filled with stones, here stem the force of the torrent brawling through its narrow bed; and some of these have held their place so long that young trees have had time to take root and flourish in them. Next comes Ospitale, another little brown-roofed hamlet perched on a green rise like Castel Lavazzo, with the usual cluster of saw-mills and saw-pits down by the water's edge; and now, entering the commune of Perarolo in a smart shower, we rattle through a succession of tiny villages built in the Swiss way, with wooden balconies, outer staircases, and deep projecting eaves. In most of these places, it being now between ten and eleven o'clock A.M., the good people are sitting in their doorways dining primitively out of wooden bowls.

So we go on; and so the Piave, greenish grey in colour, interrupted by a thousand rapids, noisy, eager, headlong, comes ever rushing towards us, and past us, and away to the sea. So, too, the brown and golden pine-trunks come whirling down with the stream. It is curious to watch them in their course. Some come singly, some in crowds. Some blunder along sideways in a stupid, buffeted, bewildered way. Some plunge madly up and down. Some run races. Some get tired, rest awhile under shelter of the bank, and then, with a rouse and a shake, dash back again into the throng. Others creep into little stony shallows, and there go to sleep for days and weeks together; while others, again, push straight ahead, nose first, as if they knew what they were about, and were bent on getting to their journey's end as quickly as possible.

Nearing Perarolo, glimpses of the peaks, aiguilles and snow-fields of Monte Cridola (8,474 feet), the highest point of the Premaggiore range, are now and then seen to the right, through openings in the lower mountains. Monte Zucco abruptly blocks the end of the gorge. Country carts upon the road, women working in the fields, a party of children scrambling and shouting among the bushes by the wayside, now indicate that we are not far from a more thickly inhabited place than any of the preceding villages. Then the road takes a sudden turn, and Perarolo, with its handsome new church, new stone bridge, public fountain, extensive wood-yards, and general air of solid prosperity, comes into view.

Yet a few yards farther, and a second bridge is crossed—a new valley rich in wood and water opens away to the left—and a wonderful majestic vision, draped in vapours and hooded in clouds, stands suddenly before us!

The coachman, preparing his accustomed *coup de théâtre*, is not allowed to speak. We know at once in what Presence we are. We know at once that yonder vague and shadowy mass which soars beyond our sight and seems to gather up the slopes of the valley as a robe, can be none other than the Antelao.

A grand, but a momentary sight! The coachman, with a jealous glance at the open maps and guide-books that have forestalled his information, whips on his horses, and in another moment valley and mountain are lost in the turn of the road, and we are fast climbing the hill leading to the great zigzag of Monte Zucco. Still we have seen, however imperfectly, the loftiest of all the giants of Cadore; we have seen the mouth of the famous Ampezzo Thal, and we begin to feel that it is not all a dream, but that we are among the Dolomites at last.

And now, for a weary while, partly on foot and partly in the carriage, we toil on and on, up the new road constructed of late years by the Emperor Ferdinand. The Piave, here quite choked by a huge, stationary mass of pine-trunks, winds unheard some hundreds of feet below. Perarolo, the great centre of all this timber trade, dwindles to a toy hamlet in the valley. New peaks rise on the horizon. New valleys glitter in the distance. Still the road climbs—winds among vast slopes of pine-forest—makes the entire circuit of Monte Zucco, and finally, with one long, last pull, reaches the level of the upper plateau.

Here, at Tai Cadore, a tiny village backed by culti-vated slopes, we are to take our midday rest. Here, too, we catch our first glimpse of Titian's birthplace, Pieve di Cadore, a small white hamlet nestled in a fold of the hills close under a ruined castle on a wooded knoll, about a mile away. Now Pieve di Cadore was down in our route as a special excursion to be taken hereafter from Cortina in the Ampezzo valley; but our impatience was great, and the sun was shining brilliantly, and our first thought was to employ these two hours' rest in walking there and back, and just seeing (though it were only the outside of it) the house in which the great painter was born.

It was first necessary, however, to take luncheon at Tai; which we did, seated at a bare deal table in an upper room of the clean little inn, beside a window commanding a magnificent view of the Premaggiore range. Meanwhile the capricious sky clouded over again; and by the time we should have been ready to start, the rain was coming down so heavily that Pieve di Cadore was unavoidably left to be seen later on.

A little way beyond Tai Cadore begins one of the finest drives in Europe. The road enters the Ampezzo Thal at an elevation which can scarcely be less than 1,250 feet above the foaming Boita; and a close, lofty, richly wooded valley, like a sublimer Val d'Anzasca, opens the way to more rugged scenery beyond. Vast precipices tower above; scattered villages cling to the green slopes half way down; and brilliant passages of light and shadow move rapidly over all. Now one peak is lighted up, and now another. Here a brown roof, wet from the last shower, glistens like silver in the sunshine; there a grassy slope fringed with noble chestnuts glows in a green and golden light; while on yonder opposite height, a dark fir-forest shows blue and purple in angry storm shadow.

At Venas, the overhanging eaves, outer staircases, and balustraded balconies, are wholly Swiss; while inscriptions such as "Qui si vende Vino d'Asti, Coloniale, ed altri generi," remind us that, although close upon the Austrian frontier, we are not yet out of Italy.

And now the valley widens. The Antelao, still

obscured by floating mists, again comes into sight—a near mass of clustered pinnacles; then the Pelmo on the opposite side of the valley, uplifted in the likeness of a mighty throne canopied by clouds, and approached by a giant staircase, each step of which is a precipice laden with eternal snow and trodden only by the chamois hunter; next, on the same side as the Pelmo but farther up the valley, appears the Rochetta—a chain of wild confused crags, like a line of broken battlements, piled high on huge buttresses of sward and pine-forest.

Between the small wayside hamlets of Vodo and Borca, the road is cut through an enormous slope of stony débris, the scene of a bergfall which fell from the Antelao in 1816, and overwhelmed two villages on the opposite bank of the Boita. More sudden, and almost more cruel than the lava from Vesuvius, it came down, as almost every bergfall comes down, at dead of night, crushing the sleepers in their beds and leaving not a moment for escape.

Two great mounds of shattered limestone, each at least 100 feet in height, mark the site of the lost villages; and, strange to tell, the torrent, instead of being dammed and driven back as at Serravalle, flows on its way unimpeded save by a few Titanic boulders. How so tremendous a fall could have crossed the stream in sufficient volume to bury every house, church and campanile on the other side, and yet have failed to fill up the bed of the intervening torrent, is infinitely mysterious. I inquired then and later whether the stream might not have been temporarily choked, and afterwards cleared by the labour of the other Ampezzan communities; but though all whom I asked seemed to think such a task impossible of fulfilment at any time, none could answer me.

"It happened, Signora, fifty-six years ago," was the invariable answer. "Chi lo sa?"

Was that so long a time? It seemed strange that, after the lapse of little more than half a century, every detail of so terrible a catastrophe should be forgotten in a place where events were necessarily few.

And now, following the great sweep of the road, we make at least one-third of the circuit of the Antelao, which becomes momentarily grander, and changes its aspect and outline with every turn. The snow on this side finds no resting place, save on a scant ledge here and there; and the mountain consists apparently of innumerable jagged buttresses, huge slopes of shaley débris, and an infinitely varied chain of pallid peaks and pinnacles. Some of these are almost white; some of a pale sulphurous yellow streaked with violet; some splashed with a vivid, rusty red, indicating the presence of iron. One keen, splintered aiguille, sharp as a lance and curved as a shark's tooth, looked like a scimitar freshly dipped in blood.

Now, at San Vito, the Antelao begins to be left behind, and the long ridge of the Croda Malcora, with its highest peak, Sorapis, standing boldly out against a background of storm-cloud, enters on the scene. A little farther yet, and the Austrian frontier is reached.*

* Chiapuzza is the last Italian hamlet, and Acquabuona is the first Austro-Tyrolean village. San Vito is also called Borea. (*Note to Second Edition.*)

A striped pole, alternately black and yellow, like a leg of one of the Pope's guard, bestrides the road in front of a dilapidated little custom-house. Here some three or four ragged-looking Austrian soldiers are playing at bowls, while a couple of officers lounging on a bench outside the door, smoke their cigarettes and watch the game. One of these, very tall, very shabby, very dirty, with a glass screwed into his eye and a moustache about eighteen inches in length, saunters up to the carriage door. Being assured, however, that we carry nothing contraband, he lifts his cap with an indescribable air of fashionable languor, and bids the coachman drive on.

From this point, the invisible political line being passed, one observes an immediate change not only in the costumes, but in the build and features of the people. They are a taller, fairer, finer race. The men wear rude capes of undressed skins. The women (no longer bare-legged, no longer *coiffées* with red and yellow handkerchiefs) wear a kind of Bernese dress consisting of a black petticoat, a black cloth bodice like a tightly fitting waistcoat, white linen undersleeves reaching to the elbow, a large blue apron, and a round felt hat, like a man's.

By this time the Pelmo is out of sight, the Rochetta is left behind, Sorapis is passed, and still new mountains rise against the horizon. To the left—a continuation, indeed, of the Rochetta—the Bec di Mezzodi and the ridge of Beccolungo stand out like a row of jagged teeth. On a line with these, but at least a mile farther up the valley, the huge bulk of the Tofana looms up in sullen majesty, headed by a magnificent precipice, like a pyramid of red granite; while to the right, Monte Cristallo, a stupendous chevaux de frise of grey and orange pinnacles, forms a grand background to the clustered roofs, lofty campanile, and green pasturages of Cortina.

For at last we are in sight of the place which is to be our head-quarters for the next week, and the wonderful drive is nearly at an end. Already, within the compass of some fifteen English miles (*i.e.*, from Tai to Cortina), we have seen six of the most famous Dolomites, three on the right bank and three on the left of the Boita. Four out of the six exceed 10,500 feet in height; while the Antelao * is, I believe, distanced by only two of its rivals, namely, the Marmolata and the Cimon della Pala. The new and amazing forms of these colossal mountains; their strange colouring; the mystery of their formation; the singularity of their relative positions, each being so near its neighbour, yet in itself so distinct and isolated; the curious fact that they are all so nearly of one height; their very names, so unlike the names of all other mountains, high-sounding, majestic, like relics of a prehistoric tongue—all these sights and facts in sudden combination confuse the imagination, and leave one bewildered at first by the variety and rapidity with which impression after impression has been charged upon the

---

* The relative altitudes of the Ampezzo Dolomites, as nearly as has yet been ascertained, are as follows :—Antelao, 10,897 feet; Sorapis, 10,798 feet; Tofana, 10,724 feet; Cristallo, 10,644 feet; Pelmo, 10,377 feet; and La Rochetta, 7,793 feet.

memory. It was therefore almost with a sense of relief that, weary with wonder and admiration, we found ourselves approaching the end of the day's journey.

HIGH STREET, CORTINA.

And now the road, which has been gradually descending for many miles, enters Cortina at about a hundred feet above the level of the Boita. First comes a scattered house or two—then a glimpse of the old church, the cemetery, and the public shooting-ground,

E

in a hollow down near the river—then a long irregular street of detached homesteads, hostelries, and humble shops—the new campanile, the pride of the village, 250 feet in height—the post-house at the corner of a little piazza containing a public fountain—and finally, being the last house in the place, the Aquila Nera,* a big substantial albergo built in true Tyrolean fashion, like a colossal Noah's ark, with rows upon rows of square windows with bright green shutters, and a huge roof with jutting eaves that looks as if it ought to take off like a lid to let out the animals inside.

This, then, is our destination, and here we arrive towards close of day, rattling through the village and dashing up to the door with our driver's usual flourish, just as if the greys, instead of having done thirty-five miles to-day and thirty-four yesterday, were quite fresh, and only now out of the stable. The Ghedinas, a father and two sons, come out, not with much alacrity, to bid us welcome. The writer, however, mentions a name of might—the name of Francis Fox Tuckett; and behold! it acts upon the sullen trio like a talisman. Their goodwill breaks forth in a ludicrous medley of Italian and German. How! the Signora is a friend of "Il Tuckett"—of the "gran' brave Signore" whose achievements are famed throughout all these valleys? Gott in Himmel! shall not the whole house be at her disposal?

---

* At the time when the First Edition of this book was issued, the Aquila Nera and the Stella d'Oro were the only hotels in Cortina; these are now much enlarged, and at least two new ones—the Croce Bianca and the Ancora —have been opened. There is English Church service now at the Aquila Nera, and old Ghedina is dead. (*Note to Second Edition.*)

Ecco! the Aquila Nera will justify the recommendation of " il brave Tuckett ! "

Hereupon we alight. The old landlord puts out an enormous brown paw; we shake hands all round; the Kellnerin is summoned; the best rooms are assigned to us; the cooks (and there seem to be plenty of them in the huge gloomy kitchen) are set to work to prepare supper; a table is laid for us on the landing, which, as we find henceforth, is the place of honour in every inn throughout the Dolomite Tyrol; and all that the Aquila Nera contains is laid under contribution for our benefit.

It is a thorough Tyrolean hostelry, by no means scrupulously clean, yet better provided and more spacious than one would have expected to find even in this, the most important village of the district. The bedrooms are immense, though scantily furnished. A few small mats of wolf and chamois skins are laid about here and there; but there is not such a thing as a carpet in the house. At the Dépendance, however—a new building on the opposite side of the road, charmingly decorated with external frescoes by one of the younger Ghedinas, who is an artist in Venice—there are smaller rooms to be had, with good iron bedsteads and some few modern comforts. But we knew nothing of this till a day or two after, when we were glad to move into the more quiet house, though at the cost of having always to cross over for meals.

In the way of food, a kind of rough plenty reigns. Luxuries, of course, are out of the question; but of veal, sausage, eggs, cheese, and sauer-kraut there is abundance. Drovers, guides, peasant-farmers and travel-

E 2

lers of all grades are eating, drinking, smoking, all day long in the public rooms, of which there are at least four in the lower floors of the big house. The kitchen chimney is smoking, the cooks are cooking, the taps are running " from morn till dewy eve." We, arriving at dewy eve, come in for an all-pervading atmosphere of tobacco and garlic—the accumulated incense of the day's sacrifices.

With all this plenty, however, and all this custom, the wealthiest and most fastidious traveller must fare off the same meats and drinks as the poorest. The only foreign wine that Ghedina keeps in his cellar is a rough Piedmontese vintage called Vino Barbera, which costs about two francs the bottle. If you do not like that, you must drink beer ; or thin country wine, either red or white ; or an inexpressibly nauseous spirit distilled from the root of a small plant nearly resembling the ordinary *Plantago major*, or common English plantain. An inferior kind of Kirschwasser is, I believe, also to be had ; but as for brandy, I doubt if there is one drop to be found in the whole country between Belluno and Bruneck.

For the rest, the inn is well enough, though one feels the want of a mistress in the establishment. Ghedina *père* is a wealthy widower, and his three stalwart sons, all unmarried, live at home and attend, in a grim unwilling way, to the housekeeping and stabling. Their horses, by the way, are first-rate—far too good for rough country work ; while in the adjoining outbuildings are to be found a capital landau, a light chaise, some three or four carettini, and——a side saddle ! How this article, in itself neither rare nor beautiful, came presently to occupy the foremost place in our affections and desires; how we fought for its possession against all comers; how we begged it, borrowed it, and finally stole it, will be seen hereafter.

Meanwhile, arriving late and tired, we were glad to accept the big rooms in the big house; to put up with the atmosphere; to sup on the landing; to hear the downstairs revellers going away long after we were in bed; and even to be waked by the wild cry of the village watchman at intervals all through the dark hours of the night. It was not, perhaps, quite so agreeable to be aroused next morning at earliest dawn by a legion of carpenters in the street below flinging down loads of heavy planks, driving in posts by the wayside, hammering, shouting, and making noise enough to wake not only the living but the dead. For this, however, as for every discomfort, there was compensation at hand; and our satisfaction was great on being told that the grand yearly Sagro, or church-festival, would be celebrated a few days hence, and that our noisy friends outside were already beginning to erect booths in preparation for the annual fair which is held at the same time. It is the most important fair in all this part of the Austrian and Italian Tyrol, and is attended by an average concourse of from twelve to fifteen hundred peasants from every hill and valley for nearly thirty miles round about Cortina.

AT CORTINA.

CORTINA, ITS SITUATION, CLIMATE, AND TRADE—A MESSA CANTATA—
THE VILLAGE CEMETERY—A FIRST ASCENT—THE GHEDINAS AND
THEIR ART—AN UNKNOWN MOUNTAIN—AN AFTERNOON STROLL—
THE ANTELAO—PLEASANT TYROLEAN WAYS—STROLLING ACROBATS—
DISSOLUTION OF PARTNERSHIP BETWEEN OURSELVES AND THE
COURIER—DIFFICULTIES ARISING THEREFROM—SANTO SIORPAES—
THE SIDE-SADDLE QUESTION AGAIN—A TYROLEAN "CARETTA"—
NEAR VIEW OF THE TOFANA—AMAZING COSTUMES—THE PEZZÉS—
SUMMIT OF THE TRE SASSI PASS—THE MARMOLATA—THE "SIGNORA
CUOCA."

CHAPTER IV.

AT CORTINA.

SITUATE on the left bank of the Boita which here runs nearly due north and south, with the Tre Croci pass opening away behind the town to the east, and the Tre Sassi Pass widening before it to the west, Cortina lies in a comparatively open space between four great mountains, and is therefore less liable to danger from bergfalls than any other village not only in the Val d'Ampezzo but in the whole adjacent district. For the same reason, it is cooler in summer than either Caprile, Agordo, Primiero, or Predazzo; all of which, though more central as stopping places and in many respects more convenient, are yet somewhat too closely hemmed in by surrounding heights.

The climate of Cortina is temperate throughout the year. Ball gives the village an elevation of 4048 feet above the level of the sea; and one of the parish priests—an intelligent old man who has devoted many years of his life to collecting the flora of the Ampezzo—assured me that he had never known the thermometer drop so low as fifteen degrees * of frost

* Reaumur.

in even the coldest winters. The soil, for all this, has a bleak and barren look; the maize (here called *grano Turco*) is cultivated, but does not flourish; and the vine is unknown. But then agriculture is not a speciality of the Ampezzo Thal, and the wealth of Cortina is derived essentially from its pasture-lands and forests. These last, in consequence of the increased and increasing value of timber, have been lavishly cut down of late years by the Commune—too probably at the expense of the future interests of Cortina. For the present, however, every inn, homestead, and public building bespeaks prosperity. The inhabitants are well-fed and well-dressed. Their fairs and festivals are the most considerable in all the South Eastern Tyrol; their principal church is the largest this side of St. Ulrich; and their new gothic Campanile, 250 feet high, might suitably adorn the piazza of such cities as Bergamo or Belluno. The village contains about 700 souls, but the population of the Commune numbers over 2500. Of these, the greater part, old and young, rich and poor, men, women and children, are engaged in the timber trade. Some cut the wood; some transport it. The wealthy convey it on trucks drawn by fine horses which, however, are cruelly overworked. The poor harness themselves six or eight in a team, men, women, and boys together, and so, under the burning summer sun, drag loads that look as if they might be too much for an elephant. Going out, as usual, before breakfast the morning of the day following our arrival at Cortina, the first sight that met my eyes was a

very old woman, perhaps eighty years of age, and a
sickly little boy of about ten, roped to a kind of
rough sledge piled up with at least half a ton
weight of rough planks.

Eight o'clock mass is performed at each church alternately, every morning throughout the year. To-day
it happened to be down at the old church, and thither,
attracted by their quaint costumes, I followed a party
of chattering peasant girls, some of whom had their
milk cans and market baskets in their hands. These
they carried into the church, taking off their hats at
the door, like men, and remaining uncovered throughout the service. The congregation consisted of some
three or four score of very old women with scant white
polls; a sprinkling of square-headed robust-looking
damsels with silver pins in their clubbed and plaited
hair; and a few old men, so tanned and gnarled and
bent that they looked as if carved out of rough brown
wood. Then trooped noisily in some four hundred
children of both sexes, and filled the benches next the
altar, while the old bell-ringer, having rung his last
peal, came hobbling up the aisle in heavy wooden
clogs and baggy breeches, and lit the candles on the
altar. Presently appeared a priest in black and gold
vestments, attended by a little red-headed acolyte, like
one of John Bellini's angels; the organist (by no
means a bad player) led off with "Ah che la Morte" on
a tremolo stop; the congregation dropped on their
knees; and the service began.

Musically speaking, it was one of those performances
which one enjoys the more the less one hears of it. A

showy operatic mass by some modern Italian composer, a reedy organ, and a choir which might have been better trained, made up an *ensemble* that soon sent the writer creeping towards the door.

It was delightful to get out again into the glorious morning. The sun was now shining deliciously; the air was heavy with the scent of new-mown hay; and the birds were singing their own little Hymn of Praise in a way that turned the Cortina choir to unmitigated discord. It was one of those mornings steeped in dewy freshness, when distant sounds and sights are brought supernaturally near, when lights are strangely bright, and shadows transparent, and the very mountains look more awake than usual. Even Tofana, rarely seen without a turban of storm-cloud, rose sharp and clear to-day against the sky.

Just opposite the old church lies the village cemetery. The gate stood ajar, and I went in—not certainly expecting to find the " God's Acre " of this wealthy commune a mere weedgrown wilderness. But so it was. Here a confusion of rough stone-heaps marking the graves of the poor—yonder a few marble tablets and iron crosses against the wall, recording the names of the better-class dead—everywhere coarse deep grass, thistles, nettles, loose stones, broken pottery and trampled clay. A couple of hand-biers, a pile of black tressels, a spade and a coil of rope, lay ready for use under a stone arcade at the farther end of the en-closure. Not a flower was there, not a touch of poetry or pathos in the place; nothing but indifference, irreve-rence, and neglect. This ugly sight, somehow, brought

back the recollection of an alms-box that I had seen
not long ago outside a pretty little cemetery near
Luino, bearing the following inscription:—" Messe
Funerale. Nel nome della Beata Maria, carità per
noi." (Funeral Masses. We implore charity in the
name of the Blessed Mary.) This appeal, coming like
a voice from the dead, had struck me at the time as
very awful; but here it would have been still more
awful, and more appropriate.

Going homewards, I found sheds and booths of all
sizes springing up the whole length of the village street,
and a great wooden enclosure like a circus being
erected in the piazza opposite the albergo of the Stella
d'Oro. A huge coloured poster, representing feats of
the trapeze, clowns, human pyramids and the like,
pasted on a space of blank wall close by, sufficiently
accounted for the shape and size of this building.

" But what is the Sagro?" I asked of a young priest
who was gravely watching the carpenters at their work.
" Is it a fair? "

" It is a festival of the Church, Signora," he replied,
with an air of reproof, and walked away.

A Sagro, however, as I soon came to know, is both
a fair and a religious festival; and it takes place once a
year in every village on the anniversary of the conse-
cration of the church, or on the festa of the saint to
whom the church is dedicated. And there are so
many villages scattered about the country, that a
Sagro is said to be going on somewhere every day in
the year.

Hurrying back now to breakfast, I found the Ghedinas, our courier, and a group of guides and peasants assembled outside the door of the Aquila Nera, staring up at the rugged peak known as the Bec di Mezzodi, on the opposite side of the valley. Telescopes were being passed from hand to hand, amid exclamations of " Eccoli ! " " Brave Signore ! " " Brave Inglese ! " —and old Ghedina, steadying his own glass for me against an angle of wall, bade me look " up yonder " for my countrymen.

Two English gentlemen then staying with their wives in the Dependance of the Aquila Nera had, it seemed, this morning achieved the first ascent of that singular peak so aptly described by Mr. Gilbert as a " carious tooth of Dolomite." The Bec itself looked neither very high nor very difficult, but I afterwards learned that it was peculiarly steep and fissured, and that they had hard work to conquer it. Ghedina's glass proved to be a good one, and I distinctly saw the figures of the climbers and their guides standing together on the topmost peak, relieved against the sky.

It being our intention to spend some little time at Cortina, thence making such excursions as lay within easy reach, we decided to devote this first day to getting ourselves acquainted with the general " lay " of the country. The most effectual way of achieving this end is, of course, to ascend some height; so, having consulted Ghedina's written list of excursions, we agreed to spend the morning in rambling about the village, and after luncheon to stroll up to the Crepa di Belvedere—a little summer house, or Jäger-lodge, lately erected at a point of view on the face of a cliff overlooking Cortina and the valley, about an hour and a half's easy walk from the village, and about twenty minutes to the left of the cross on the road to the Tre Sassi Pass. The Belvedere, a tiny white speck against a scar of red cliff in the midst of a long sweep of fir-forest, is seen from the windows of the inn and lies before the climber all the way.

Meanwhile, however, we breakfasted, wrote letters, examined the paintings and frescoes in and about the two houses, and made arrangements for shifting our quarters into the quieter and better furnished rooms over the way. Two of the younger Ghedinas, it seemed, were painters; a third carved cleverly in wood; and the fourth (a grave practical man devoted to the business, the stabling, and the wood trade) played a trombone in the village band.

Both houses are full of heads and studies in oil, designs for large pictures, and sketches of unequal merit. A head of a bearded man in one of the upper chambers of the Aquila Nera, and two half-lengths of his father and mother in the dining room, may be taken as fair specimens of the skill of the portrait-painting son; while the external frescoes of the Dependance, two in the new church, and all sorts of rough and ready designs, some military, some religious, some grotesque, flung here and there upon the walls of stair-cases, cart-sheds, neighbours' house-fronts and so forth, represent the superior gifts and culture of the brother who lives in Venice. As for the decorations of the Dependance, they are full of power, and to the sound drawing and skilful designing of the Munich school, add a warmth and tenderness of colour almost Italian. Three large groups representing Sculpture and Architecture, Painting, and the Physical Sciences, and three medallions containing portraits of Raffaelle, Titian and Albert Dürer, cover all that is not window-space above the ground floor. The figure of Mercury in the first group and of Urania in the last, and the way in which such stubborn objects as the steam-engine, camera, and telegraphic apparatus have been pictorially treated, are deserving of particular notice. To Albert. Dürer, like a true German, the artist gives the middle place among the medallions.

Very different, though almost as good in their way, are the mounted Cossacks, wild horses, and mediæval men-at-arms that skirmish all over the whitewashed walls of the outhouses and stables of the Aquila Nera; to say nothing of the fantastic devil, all teeth and claws, that grins upon unsuspecting customers from outside the stove in the only chemist's shop in Cortina. We asked for the painter; but he was far away in Venice, and his studio, they told us, was not only closed but empty.

To ascend the Campanile and get the near view over the village, was obviously one of the first duties of a visitor; so, finding the door open and the old bellringer inside, we mounted laboriously to the top—nearly a hundred feet higher than the leaning tower of Pisa. Standing here upon the outer gallery above the level of the great bells, we had the village and valley at our feet. The panorama, though it included little which we had not seen already, was fine all round, and served to impress the main landmarks upon our memory. The Ampezzo Thal opened away to North and South, and the twin passes of the Tre Croci and Tre Sassi intersected it to East and West. When we had fixed in our minds the fact that Landro and Bruneck lay out to the

UNKNOWN MOUNTAINS NEAR CORTINA.

North, and Perarolo to the South ; that Auronzo was to be found somewhere on the other side of the Tre Croci ; and that to arrive at Caprile it was necessary to go over the Tre Sassi, we had gained something in the way of definite topography. The Marmolata and Civetta, as we knew by our maps, were on the side of Caprile ; and the Marmarole on the side of Auronzo. The Pelmo, left behind yesterday, was peeping even now above the

F

ridge of the Rochetta ; and a group of fantastic rocks, so like the towers and bastions of a ruined castle that we took them at first sight for the remains of some mediæval stronghold, marked the summit of the Tre Sassi to the West.

" But what mountain is that far away to the South ?" we asked, pointing in the direction of Perarolo.

" Which mountain, Signora ? "

" That one yonder, like a cathedral front with two towers."

The old bellringer shaded his eyes with one trembling hand, and peered down the valley.

" Eh," he said, " it is some mountain on the Italian side " (E una montagna della parte d'Italia).

" But what is it called ? "

" Eh," he repeated, with a puzzled look, " chi lo sa ? *I don't know that I ever noticed it before.*"

Now it was a very singular mountain—one of the most singular and the most striking that we saw throughout the tour. It was exactly like the front of Notre Dame, with one slender aiguille, like a flagstaff, shooting up from the top of one of its battlemented towers. It was conspicuous from most points on the left bank of the Boita ; but the best view, as I soon after discovered, was from the rising ground behind Cortina, going up through the fields in the direction of the Begontina torrent. From thence I made the accompanying sketch ; and to this spot we returned again and again, fascinated as much, perhaps, by the mystery in which it was enveloped, as by the majestic outline of this unknown mountain, to which, for want of a better, we

MONTE ANTELAO.

gave the name of Notre Dame. For the old bellringer was not alone in his ignorance. Ask whom we would, we invariably received the same vague reply—it was a mountain " della parte d'Italia." They knew no more ; and some, like our friend of the Campanile, had evidently " not noticed it before."

What with the great heat of the afternoon, which made uphill work difficult and rapid walking impossible ; what with the wonderful wild flowers that enticed us continually from the path ; what with chatting to peasants by the way, stopping to study the landscape, sketching, and so forth, we never reached the chalet of the Belvedere, after all. We came very near it, however, and gained a magnificent view over the valley, the Cristallo group, and the range of the Croda Malcora. Hence also, from a grassy knoll near the cross below the Crepa, the writer devoted a long hour to making a careful drawing of the Antelao which is here seen to its greatest advantage. From no other point, indeed, is it possible, so far as I am aware, to get so good a view of the great snow slope at the back of the summit in combination with the splintered buttresses that strike down towards Borca and Vodo in the front.*

The first ascent of the highest peak of this mountain was achieved by that famous climber, Dr. Grohmann, in 1863 ; and the second, in 1864, by Lord Francis Douglas of hapless memory, accompanied by Mr. F. L. Latham and by two guides named Matteo Ossi and Santo Siorpaes. The latter—a brave, hardy, faithful

* The height of the Antelao, as determined by the last Austrian survey, is 3,320 mètres, or 10,897 English feet. (*Note to Second Edition.*)

fellow, who travelled with us later in the autumn among the Italian Alps and through the Zermatt district — assured me that Lord Francis, though so young, was an excellent mountaineer, and described him as " buono, bello, e biondino " (good, handsome, and fair).

The ascent is taken from a pass called the Forcella Piccola which divides the mass of the Marmarole from that of the Antelao, and is most quickly reached from San Vito. Owing to the long snow-slope before mentioned, this mountain, up to a certain point, is considered to be easier than any other great Dolomite except the Marmolata; but the last pull up the actual pinnacle, which rises "with formidable steepness" to a height of some three hundred feet, and curves over like a horn, is said to be difficult. It was supposed to be inaccessible till Dr. Grohmann's time, when the fortunate discovery of a certain cleft by one of his Cortina guides, opened the way to the German cragsman and to all who should come after him. A good climber can ascend from, and return to San Vito in eleven hours, exclusive of halts.

The country folk were all coming up to their homes on the pasturages of Monte Averau, as we went down again in the cool of the early evening—some with empty milk-pails, having sold their milk in Cortina; others carrying home their store of bread and flour, just purchased. One or two begged somewhat abjectly for a soldo " per l'amor di Dio; " but for the most part they passed with a brisk step, a pleasant smile, and a cheerful " Guten Abend," or " buona sera." A civil, kindly people on the whole, as we soon came to know right well! A people ready with good wishes and little friendly salutations which, even if they have come to be spoken as mere matters of course, yet help to keep warm the spirit of good will. If they pass through the room where you are at meals, they wish you " good appetite ; " if you are going out, " a pleasant walk ; " if on your way to bed, " sound sleep and happy dreams." You yawn, and they wish you " felicità ; " you sneeze, and they say " salute."

That evening, as we were sitting down to a meal which was dinner, or supper, or both, we were startled by a furious discord of drums and brass instruments in the street below. It was the company of strolling acrobats who had just arrived and were parading through the village, followed by all the boys and idlers in the place :—a drummer on stilts ; a buffoon in high collars and a tall hat, like Paul Pry ; some half dozen athletic fellows in the traditional fillets and fleshings ; and about as many hideous-looking, muscular women, tramping the dusty road in white shoes and the briefest conceivable skirts. The " theatre " it seemed was to open to-morrow, although the Sagro would not be held till Sunday.

It was on the morning of the third day after we had settled down at Cortina, that the storm which had so long been gathering, burst at last. Supported by the consciousness of his own merit, the courier had borne with us till he could bear with us no longer. Now, however, the near prospect of being dragged over passes and up mountains, of having to ride on a mule for days in succession, and of living for many weeks to come in Tyrolean albergos several degrees less comfortable than the Aquila Nera, was too much for the great man's philosophy. He understood, he said, that there were no carriage-roads to most of the places laid down in our maps, and " no suitable accommodation such as he was accustomed to when travelling with parties who placed confidence in his opinion ; " he therefore begged leave to tender his resignation, and his accounts. Our vaga-bond tastes, in short, were too much for him ; and he deserted us (if that could be called desertion which must in all likelihood have taken the form of dismissal ere long) just at the time when the protection of a trust-worthy and respectable man had become an indispensa-ble condition of our journey.

It is needless to add that the fortnight's notice which he offered was summarily rejected, and that he was then and there paid off and done with. As for L., by whom he had been retained for months before we joined forces in Naples, she transacted the whole affair with an amount of withering *sang-froid* which speedily reduced the offender to a condition of abject humility. He made an effort by-and-by to assert his indifference by playing at bowls in front of the albergo ; but went away in the afternoon outside the Longarone Stell-wagen, quite crestfallen.

And now, what was to be done? Could we possibly go on with only guides, and no courier? Or must the tour through the wild heart of the country be given up, just as we had come within sight of our promised land? These were questions that must be solved before we could venture one day's journey beyond the post roads of Cortina.

As a matter of choice, we infinitely preferred the absence of our discontented friend. It was so delicious, indeed, to be without him, that L. said she felt as if a necklace of millstones had been taken from round her neck; but then, as a matter of expediency, his defection was undeniably inconvenient. Could he, however, be in any way replaced—not, of course by another courier, that kind of article being quite unknown in these primitive valleys; but by some reliable man, as, for instance, Santo Siorpaes, who had been especially recommended to us beforehand, and who was reputed to be the best head-guide in Cortina?

To send for him and offer him an engagement for the whole journey was the first step to be taken. He came:—a bright-eyed, black-haired mountaineer about forty; a mighty chamois hunter; an ex-soldier in the Austrian army, and now a custode of forests, and local inspector of roads; an active, eager fellow, brown as a berry, with honesty written in his face, and an open vivacious manner that won our liking at first sight.

Unfortunately, however, this jewel of a guide was pledged for the next six or eight weeks and could not by any means get free. Had he no friend, we asked, whom he could recommend to take his place? He pondered the question, and looked doubtful. There was old Lacedelli, he said, but he was too old; and there was young Lacedelli, but he was too young. Also there was a certain Angelo, but he was away, and would not be back for a month. Then, again, most of the men about Cortina were good enough at rough climbing, but not used to travelling with ladies. Well, he would think it over—he would think it over, and let the Signoras know. But when would he let us know? This evening? He shook his head. This evening he was engaged to start for some distant valley with a party of gentlemen who were to ascend a mountain to-morrow. No—he could not promise to see us again before Sunday; but he would then wait upon us after High Mass.

This was all we could obtain from him. It was not much; and we began to have dismal forebodings of the failure of our plans.

Meanwhile, however, it was of no use to despond. There was plenty to be done at Cortina, whatever happened. We could go to Pieve di Cadore, to Auronzo, to Landro, by good carriage roads. We could see about the side-saddles. We could even go in what our landlord called a " caretta " as far as Falzarego, the hospice on the summit of the Tre Sassi pass, and thence obtain a view of the Marmolata.

During the present uncertainty, it was some comfort first of all to agitate this question of the side-saddles. In the event of our being able to carry out the journey, they were of more real importance than a whole army of couriers. Without them, certainly, we could do nothing in the way of peaks or passes.

Now we knew from previous information that Madame Pezzé, landlady of the inn at Caprile, had a saddle which was presented to her for the use of lady travellers by F. F. T. A persuasive note couched in the writer's best Italian was therefore sent over by a special messenger, who had instructions to bring the precious object back, if possible, upon his shoulders.

Then old Ghedina also possessed one; but, divining perhaps that we should be over-long borrowers, was particularly reluctant to show it. It was not till the writer succeeded in following him one day into the stable, that this mysterious treasure was allowed to see the light. It proved to be a fairly good saddle; but then it was only one, and if we even obtained Madame Pezzé's, we should still require a third.

"I am expecting a new *sella di donna* from Vienna," sputtered the old landlord, in his polyglot patois. "*Ein schöner Sattel!*"

"When will it arrive?" I asked eagerly.

"*Diavolo!* I don't know. Perhaps to-night—perhaps next week. I have been expecting it every day for the last three months!"

I relapsed into hopelessness.

The old man grinned from ear to ear—he had a large, brown, flat face that looked as if it had been sat upon—and patted me on the shoulder with a paw like a Bengal tiger's.

"Tut! tut!" he said, "you are a brava Signora—you shall not be disappointed. We'll dress up a *Basta* for the *cameriera*, and all shall be well!"

This promise of the Basta was obscure, but comforting. I had not the slightest idea of what a Basta was, and Ghedina could only tell me what it was not. It was not a side-saddle. It was not a chair. It was not a railed seat with a foot-rest, like a child's donkey saddle.

It had to be made when required, and should be forth-coming when wanted. Beyond this point we could not get; and there the matter had to rest, at all events for the present.

Next morning we ordered the caretta to take us to Falzarego. It would be difficult, perhaps, to say why, but we were longing to see the Marmolata, and could not rest till we had achieved, at least, a distant glimpse of him. In the first place, it is supposed to be the highest of all the Dolomites; in the second, its snowfields and glaciers are more extensive than those of any of its neighbours; and in the third place, it is so hemmed in on all sides by other mountains that it is very difficult to obtain a view of it at all.*

The morning was somewhat doubtful. The Tofana had on its helmet of cloud, and though the sun shone brilliantly at times, there was an unsettled, uncertain look about the rolling cumuli that kept us hesitating till nearly eleven A.M. Then old Ghedina pronounced in favour of the weather, and we resolved to venture.

I shall not soon forget our dismay at first sight of the caretta. It was simply a wooden trough on four wheels, some seven feet in length by three and a half in breadth, with a cross-wise plank to sit upon. The horse—a magnificent light chesnut full seventeen hands high, with a huge leather collar like an Elizabethan ruff—towered above the vehicle; and a boy sat on the shafts to drive. Springs, of course, there were none; cushions

---

* Except from some considerable height, such as the top of the Tre Sassi the Col d'Alleghe, or the Col Fiorentino, I know few points from which it is even visible.

there were none; but mats and rugs were piled in abundantly, and so we started.

Our way lay over the bridge and up past the cross where we had rested and sketched a day or two before. Again the great view over the valley became unrolled like a scroll beneath our feet. Again the Cristallo, the Croda Malcora, Sorapis and Antelao seemed to rise as we rose, and the Tofana loomed nearer and more threatening with every step of our progress. Now, mounting ever higher among green slopes gorgeous with wild flowers, and through pine-woods all abloom with strawberry blossoms, we left the Cortina view behind, and passed close under the south-west face of the Tofana—so close that we could distinctly see the mouth of a famous cavern which is said to penetrate for many hundred feet into the heart of the mountain. Seen from the Tre Sassi road, it looks perfectly inaccessible—a mere rabbit-hole in the face of a vertical and triangular precipice, like the entrance to the Great Pyramid. This cavern, however, is one of the sights of Cortina, and can be reached without difficulty when there is an accumulation of snow upon the slopes beneath.

And now, as we mount higher, rounding the last buttresses of the Tofana and coming in sight of the first outlying ridge of Monte Lagazuoi, we begin to meet frequent groups of peasants, some two and three, some twelve or fifteen strong; some carrying huge loads of home-spun frieze and linen on their backs; some laden with wooden ware; some with live poultry; all in their holiday clothes, and all bound for the great Sagro. They are of all ages, and apparently of all grades; old folks and young, farmers and farm-servants—a stumpy, sturdy, fresh-coloured, honest-looking race; the women with legs like pillars, and the men averaging from five foot five to five foot seven in height. The old men wear knee breeches and comical little frieze coats very short and full in the skirts, with two large buttons set high up in the middle of their backs, like a pair of eyes. The young fellows affect trousers and embroidered braces, and carry little bunches of coloured feathers and artificial flowers in their hats. The costumes of the girls, however, are quite over-whelming, and unlike anything that we have yet seen. They wear hats like the men, and adorned in the same manner; dark green, blue, or brown skirts laid in close folds like the plaiting of a kilt, and starting from just between the shoulders, like a sacque; bodices open in front and laced with purple braid; sleeves tight to the arm and wrist, but slashed at the top with a puffing of white linen; and round their necks bright scarlet and yellow handkerchiefs of printed cotton.

"What people are these?" we ask, as the first of many such apparitions appears before us at a turn of the road.

To which the boy on the shafts—a laughing, merry fellow named Giovanni—replies that these are contadine from Buchenstein, Livinallungo, and Corfara.

"But Corfara is a long way off!" exclaims L.,

who is better up in her maps than myself, and knows something of the distances.

"Eh! some of them come forty, fifty, sixty miles over the mountains—some walk all night both coming and going. Ecco!" (with a critical glance at the pillars before-mentioned) "what are the miles to a donzella like that!"

Meanwhile, we are suffering agonies of dislocation; for the road (which is only just wide enough for our wheels, and overhangs a precipice at the bottom of which foams a roaring torrent) is full of loose stones, over which the caretta jolts and blunders, creaks, leaps and rolls in such a distracting manner that we are fain at last to get out and walk.

The glen now grows narrower, and the castellated rocks which we had already observed from Cortina are seen high above sloping woods on the opposite bank of the stream. Giovanni, who knows everything, informs us that they are here called the Torette, and form part of the crest of Monte Nuvolau;* and that the torrent, which takes its rise somewhere among the fastnesses of Lagazuoi, is known as the Costeana.

More and more pedestrians, meanwhile, keep trooping past. The farther we go, the thicker they come. Where will they all sleep to-night? The Aquila Nera and the Stella d'Oro, were they each four times their present size, would not hold more than half

---

* Giovanni can hardly have been right. The Nuvolau lies W.S.W. of Cortina, and would not have been visible in that direction. The "Torette" more probably belong to the Becca di Mezzodi or La Rochetta.

of them; and yet this is only one road out of many. At this moment they are tramping into Cortina from Auronzo, from Pieve di Cadore, and from all the villages of the Ampezzo Thal. There will be fifteen hundred strangers, says our driver, in Cortina to-night.

And now, quite suddenly, we come upon a better-dressed group than any we have yet met—two tall, gentlemanly-looking young men and a lady, followed by a countryman with their luggage on his back. The lady is young and pretty, with a rose in her black hair, and no bonnet. The young men lift their hats as they pass. The countryman, plodding after them, looks up with a somewhat knowing expression, and touches his cap. But what is he carrying on his back? Not their luggage, after all. A side-saddle! A large, new side-saddle, with a third pommel to screw, and a velvet-lined stirrup dangling down behind. It was our own messenger—it was Madame Pezzé's saddle!

Hearing a duet of joyful exclamations in the rear, the young lady turned round, smiling. The young men came forward, smiling also. They were Madame Pezzé's two sons, Lieutenant Cesare Pezzé, an ex-Garibaldian officer, and young Agostino Pezzé, who, with his mother, keeps the inn at Caprile. The damsel with the rose in her hair was Agostino's wife. They had come over the pass on foot, and were bound, like everyone else, for the Sagro at Cortina.

Concluding, of course, that we were on our way to

Caprile, their surprise was great that we should have left Cortina without waiting for the festival; but they were still more astonished on finding that we had come up all this way only to peep at the Marmolata and go back again.

"Shall we get a good view?" I asked, somewhat anxiously; for the clouds had been gathering gloomily during the last half hour.

They shook their heads and looked doubtful. The mists were thickening fast, they said, on the other side. We must push on at once for the top, and delay for nothing at the Hospice. The mountain was quite clear half an hour ago—but soon there would be nothing of it visible.

This opinion brought our interview to an abrupt conclusion, and, with the promise of meeting again to-morrow, sent us hurrying away towards the Hospice— a small white cottage by the roadside, about a quarter of a mile ahead.

Here we left the caretta, bade Giovanni attend to the comforts of his horse, and hastened on alone towards the top. We had but to follow the road, which swept round and across a wild slope of barren moor bounded by the crags of Lagazuoi on the one hand, and by the low-lying ridge of Monte Nuvolau on the other. Tall posts, each the stem of a stout fir-tree, were here set at regular intervals along the side of the path, like telegraph posts, to mark the course of the road;—a necessary precaution at this height (7,073 feet) where the snow lies deep for eight months out of every twelve. Even now, on the sixth of July, every rift and hollow held its yet unmelted snowdrift.

And now a rough wayside cross comes into sight a few yards farther ahead—a swift runner overtakes us—and Giovanni, breathless and flushed, exclaims :—

"Ecco, Signore! Ecco la croce! Di là vedremo la Marmolata." (See, Signore! Yonder is the cross! From there we shall see the Marmolata.")

And from there, by rare good fortune, we do see it—a huge, roof-shaped mass, sloping, and smooth, and snowy white against a leaden sky. For vastness of expression and extent of snow, as seen from this side, it recalls Mont Blanc. Distance, instead of diminishing its bulk, seems by contrast with surrounding heights, to enhance it. The two valleys of Andraz and Livinallungo, the Monte Padon, and a whole sea of minor peaks occupy the intervening space ; and yet the Marmolata seems to fill the scene.

But only for a few seconds! Even as we stand there, eagerly gazing at it, the summit becomes dimmed; the outline fades ; a pale grey tint spreads over the snow-fields ; and there remains only a blurred, gigantic, indefinite Something, scarcely to be distinguished from the mists by which it is surrounded.

"Diavolo of a Marmolata!" exclaims Giovanni. "The Signoras were only just in time—but they have seen him *pulito*."

Now this word "pulito" (clean) in one sense or another, is always on the tip of Giovanni's tongue ; and, as I soon afterwards find, is used indiscriminately for clear, brilliant, successful, intelligible, and a dozen other meanings, throughout this part of the Tyrol. Your mule goes "pulito." Your new boots fit you "pulito." Your field glass shows objects "pulito." You achieve a creditable bit of climbing, and are complimented on having done it "pulito." Your driver was drunk last evening, but you are assured that he is "pulito" (in the sense of sober) this morning. It is, in short, a word of most elastic capabilities; but somewhat puzzling to strangers for that reason.

The Marmolata having retired from the scene, we now turn back, taking a short cut across the dreary "Col" and finding by the way some exquisite specimens of wild Daphne (*Daphne Cneorum*), abundance of the small mountain gentian (*Gentiana verna*), and large clusters of a very lovely, tiny pink flower * with wax-like petals, minute and close as a lichen, and unlike anything that either of us has ever seen before.

Arrived at the Hospice, and being by this time very hungry, we go in, and are welcomed by a clean, smiling padrona who (because her one public room is full of peasants eating, drinking and smoking) invites us into the kitchen—a model kitchen, like a kitchen in a Dutch picture, with a floor of bright red bricks, and a roaring wood-fire, and rows upon rows of brass and copper pans shining like mirrors. She proves to be richer, however, in cooking utensils than in provisions; for dry bread, eggs, butter, and a coarse, uneatable mountain cheese are all she has to offer.

Still, with eggs and butter one is not obliged to starve. The writer, in a moment of happy inspiration, undertakes the part of cook, and offers to concoct a

---

* Mr. Tuckett suggests that this may have been the *Androsace glacialis*.

certain dish known as "buttered eggs," or, more politely, as "hasty omelette." So an apron is borrowed, and, to the unbounded entertainment of the landlady and her servant, the savoury mess is prepared in a few minutes. From that moment I am known at Falzarego as the "Signora Cuoca" (the Signora-Cook); am greeted by that title the next time I appear at the Hospice; and am remembered by it, doubtless, to this day.

By the time we are again ready to start, the mists have rolled up to the top of the pass, and the sky all round looks black and threatening. Some peasants outside predict a storm, and counsel us to get down into the valley as quickly as may be; so the chesnut is hastily put to, and we rattle off just as the first heavy drops come splashing down to a low accompaniment of very distant thunder.

The storm, however, if there was a storm, remained locked in on the other side of the pass. We soon left it behind; and long before we reached the point leading to the Crepa di Belvedere, the sun was shining brilliantly.

CORTINA TO PIEVE DI CADORE.

THE SAGRO OF CORTINA—A TYROLEAN SERMON—THE PEASANT
MAIDEN OF LIVINALLUNGO—THE COURIER REPLACED—AN AMPEZZO
WEDDING—THE TOFANA—PEUTELSTEIN—THE HÖLLENSTEIN THAL—
THE CRODA ROSSA—LANDRO AND THE DÜRREN SEE—THE DREI
ZINNEN—THE START FOR AURONZO—THE CHURCH OF THE CRUCIFIX
—PIEVE DI CADORE—THE HOUSE IN WHICH TITIAN WAS BORN—THE
CASA ZAMPIERI—AN INVASION—TITIAN'S FIRST FRESCO—THE ODIOUS
LITTLE GIRL—THE DUOMO—DON ANTONIO DA VIA—THE CADORE
TITIANS—THE FOUR TEMPERAS—A CURIOUS ANTIQUE PREDELLA.

CHAPTER V.

CORTINA TO PIEVE DI CADORE.

THE morning of the Sagro dawned to a prodigious ringing of church-bells and firing of musketry. There were masses going on in both churches from five A.M. till mid-day. The long street and the piazza by the post-office presented one uninterrupted line of booths. There were hundreds of strangers all over the town; hundreds in the churches. Every house seemed suddenly to have become an albergo. Every window, every balcony, every doorway was crowded. The acrobats again paraded Cortina this brilliant Sunday morning about nine o'clock, and the discord of their drums and trumpets went on all day long, to the accompaniment of the church-bells and the intermittent firing of the sharp-shooters down at the " Tir " by the riverside.

What a motley crowd! What a busy, cheerful scene! What a confusion of voices, languages, music, bells and gunpowder! Here are Austrian Tyrolese from Toblach, Innichen, and the Sexten Thal, who speak only German; Italian Tyrolese from the Longarone side, who speak only Italian; others from the

border-villages who speak both, or a patois compounded
of both, which is quite unintelligible. The costumes of
these mountain-folk are still more various than their
tongues. The women of San Vito wear breastplates of
crimson or green satin banded with broad gold braid,
and ornamented with spangles. The women of the
Puster Thal walk about in huge turban-like head-
dresses, as becoming, and quite as heavy, as the bear-
skins of the Grenadiers. The men of Flitsch are lost
in their enormous black boots, modelled, apparently, on
those of the French postillion of the last century. Here,
too, are old women in home-made otter-skin hats, high
in the crown and ornamented like a footman's with a
broad gold band; and bold Jägers with wide leather
belts, green braces, steeple-crowned hats, and guns
slung across their shoulders, looking exactly like Caspar
in " Der Freischütz." The wonderful damsels of Livinallungo whom we met yesterday on the pass, are also
present in great force; but the prevailing costume is of
course that of the Ampezzo. It consists of a black felt
hat with a bunch of feathers at the side; a black cloth
skirt and bodice trimmed with black velvet or black
satin; loose white sleeves; a large blue apron that
almost meets behind; and a little coloured handkerchief round the neck. Simple, sober, and becoming,
this dress suits young and old alike; and the round hat
sets off a pretty face very agreeably.

Learning that the musical mass was to begin at
eleven A.M., we took care, as we thought, to reach the
church in good time; but at a quarter before the hour,
we found the steps crowded outside, and barely standing

room within. The whole body of the church was one mass of life, colour, bare heads and upturned faces. Men and women alike held their hats in their hands. Three priests at three different altars performed mass simultaneously. The organist played his best, assisted however, by the Cortina brass band with an effect that was almost maddening. One trombone player in particular, an apoplectic red-faced man in grey flannel shirtsleeves, blew as if bent on blowing his brains out. Now and then, when the organist had an unaccompanied interlude, or the choir-master a few phrases of solo, there came a lucid interval when one breathed again. But these respites were few and brief; and except during the sermon, the brass band that morning had quite the best of it.

The old curé preached, attired in magnificent vestments of white and gold brocade. His sermon turned upon Faith, and he illustrated his text oddly enough by references to all kinds of matters in which Faith is not generally supposed to bear a leading part. The soldier, the artist, the lawyer, the man of science, what could they do, he asked, without Faith? Take the soldier, for instance:—what is it that inspires him with courage to face the cannon's mouth? Faith. Take the painter—judge what must have inspired the frescoes and paintings in this very church:—Faith. Think of the patience and labour required in the cutting of the Suez canal! What supported those workmen through their trying task? Faith. Look again at the Mont Cenis tunnel! Think of how those engineers began at opposite sides of that great mountain, and at length, after years of labour, met in the midst of it. To what power must we attribute such perseverance crowned with such success? To the supreme and vivifying power of Faith.

Of such quality was the good man's discourse. He preached in Italian, and paused after every peroration to mop his bald head with a blue cotton pocket-handkerchief. It was a hot day, and his eloquence quite exhausted him.

Coming out of the church, we take a turn round the fair. Here are booths for the sale of everything under the sun—hats; umbrellas; pipes; spectacles; pots, pans, and kettles; tanned leather; untanned leather; baskets; wooden ladles; boots and shoes; blankets; home-spun frieze and linen; harness; scythes; tin wares; wooden wares; nails, screws and carpenters' tools; knives, forks and spoons; crockery; toys; crucifixes and prayer-books; braces, garters, pocket-books, steel chains, sleeve-buttons and stationery; live poultry; fruit; vegetables; cheap jewellery; ribbons; stuffs; seeds; bird-cages; and cotton umbrellas of many colours. Here, too, is a stall for the exclusive sale of watches, from the massive silver turnip to the flat little Geneva time-keeper of the size, and probably also of the value, of an English florin. Near the church-door stands a somewhat superior booth stocked with mediæval brass work, altar-candlesticks, patinas, chalices and the like; while, next in rotation, a grave-looking old peasant presides over a big barrel full of straw and water, round the top of which, in symmetrical array, repose whetstones of all sizes.

It is remarkable that there are here no dancing or refreshment booths. The sober Tyrolese do not often dance, unless at weddings; and for meals, those who have not brought food with them, crowd at midday into the inns and private houses, and there eat with small appearance of festivity. Even the acrobats do not seem greatly to attract them. A large crowd gathers outside the show and almost fills the piazza in the afternoon; but not many seem to be going in. They are content, for the most part, to listen to the comic dialogue sustained on the outer platform by the clown and Merry Andrew, and prefer to keep their soldi warm in their pockets.

Now the writer, knowing from previous experience the unpopularity of the sketcher, steals into corners and behind booths, in order to secure a few notes of costume and character; but, being speedily found out and surrounded, is fain either to use her pencil openly or not at all. The good people of Ampezzo, however, prove to be less sensitive in this matter than the peasants of Italy or Switzerland. They are delighted to be sketched, and come round by dozens, begging to have their portraits taken, and anxious that no detail of costume should be omitted. One very handsome woman of Livinallungo, tempted by the promise of a florin, came home with me in order that I might make a careful coloured study of her costume. She was tall, and so finely formed that not even that hideous sacque and shapeless bodice could disguise the perfection of her figure. As I placed her, so she stood, silent, motionless, absorbed, for more than half an hour. A more majestic face I never saw, nor one so full of a sweet, impenetrable melancholy. Being questioned, she said she was twenty-three years of age, and a farm-servant at Livinallungo.

" And you are not married ? " I asked.

" No, Signora."

" Nor betrothed ? "

" No, Signora."

" But that must be your own fault," I said.

She shook her head.

" Ah, no," she replied, with a slightly heightened colour. " Our young men do not marry without money. Who would think of me ? I am too poor."

I should have liked to know more of her history ; but her natural dignity and reserve were such that I felt I must not question her farther.

The sketch finished, she just glanced at it, put back the proffered payment, and turned at once to go. The Signora was very welcome, she said ; she did not wish to be paid. Being pressed, however, to take the money, she yielded, more, as it seemed, through good-breeding than from inclination ; and so went away, taking the downward path from the back of the house, and going home over the mountain, alone.

That afternoon, Santo Siorpaes came again, bringing with him a tall, brown, fair-haired young man of about twenty-eight or thirty, whom he introduced as " Signore Giuseppe Ghedina." This Giuseppe, he said, was a farmer, lately married, well-to-do, and a nephew of our landlord of the Aquila Nera. Not being a professional guide, he would nevertheless be happy to travel with the Signoras, and to be useful to the utmost of his power. He did not profess to know all the country laid down in our scheme, but he would take Santo's written instructions as to routes, inns, mules, guides and so forth; and he, Santo, did not doubt that we should find Giuseppe in all respects as well fitted for the work as himself.

Now Giuseppe's manner and appearance were particularly prepossessing. We liked his simple gravity, the intelligence with which he asked and answered questions, and the interest with which he examined our maps and guide-books. Preliminaries, therefore, were soon settled. He was to inform himself thoroughly upon all matters connected with the route, and to hold himself in readiness to join us in a day or two. Meanwhile it was agreed that we should pay him at the same rate that we should have paid Santo Siorpaes: namely two and a half florins a day for his wages, and one florin and a half for his food—in all, about eight francs, or six and eightpence English, per diem. If at any time we were to travel by any public conveyance, we were of course to pay his fare; but all lodging and other expenses *en route* were to be defrayed by himself.

It may here be observed, once and for always, that a more fortunate choice could not have been made. Faithful, honest, courteous, untiring, intelligent, Giuseppe Ghedina, unused as he was to his new office, entered upon his duties as one to the manner born, and left nothing to be desired. Always at hand, but never obtrusive, as economical of our money as of his own, he was always thinking for us and never for himself. And so anxious was he that the Signoras should see all that was to be seen, that, when travelling through a district new to himself, he used to take pains each evening to enter in his pocket-book all such details as he could pick up in advance respecting every object of interest which might chance to lie in our way in the course of the next day's journey. He remained with us, as will be seen, throughout this Dolomite tour : and we parted with mutual regret when it ended.

Numbers of those who had thronged the fair and the churches all this day, went home the same afternoon or evening. As long as daylight remained, they could be seen dotting every mountain path ; and for hours after all Cortina was in bed, their long wild Alpine cry rang from hillside to hillside, and broke the silence of the night. Next morning, however, there seemed to be as many as ever in the fair, which was kept up throughout the second day with undiminished spirit.

This second morning began with a wedding. The order of the bridal procession was as follows. First came the indefatigable brass band, numbering some twenty performers ; then the bride and the best man ; then the bride's father and mother ; then the bride-groom walking alone ; and lastly some fourteen or fifteen friends and relations of both sexes. In this order, they twice paraded the whole length of the town. The bride wore a black alpaca dress ; the usual black cloth bodice and white sleeves ; and a gorgeous apron of red and green silk fastened behind with a pair of quaint brass clasps. Neither she nor any of the other women on this occasion wore hats ; but only an abundance of silver pins in their neatly plaited hair. Having entered the church, they all took seats in the aisle about half-way down, and the band went into the organ-loft.

Presently the bridegroom went up by himself to the altar, and kneeled down. When he had knelt there a few minutes, the mother of the bride led her daughter up, placed her at his left hand, and there left her. After they had both knelt there some five minutes longer, the priest came in, followed by the old bell-ringer, who acted as clerk. The bellringer then lighted a pair of long wax tapers and handed them to the priest, who blessed them, and gave one to the bride and the other to the bridegroom. This was the beginning of the ceremony.

Then the priest read the marriage service in a low voice and very quickly, only pausing presently to ask for the rings, which were handed to him on a little glass dish by the bellringer. The priest, having blessed the rings, first gave one to the bridegroom to place upon the finger of the bride, and then gave the other to the bride, to place upon the finger of the bridegroom. During all this time they never parted from their tapers, but shifted them from one hand to the other, as occasion required. At this stage of the ceremony, the bridegroom produced some money, and gave it to the bride. They were then profusely sprinkled with holy water, and this concluded the marriage service.

High mass was next performed, as yesterday, with the full band and organ ; the newly married couple remaining the whole time upon their knees before the altar, with their lighted tapers in their hands.

At length, when all was over, and the congregation

was about to disperse, the bridegroom got up quite coolly and walked out of the church, leaving his bride still kneeling. Then her mother came up again, and led her away. The bridegroom, without so much as looking back to see what had become of her, went and played at bowls in the piazza; the bride went home with her parents, took off her finery, and shortly reappeared in her shabby, everyday clothes. It is, perhaps, Tyrolean etiquette for newly married persons to avoid each other as much as possible. At all events, the bridegroom loafed about with the men, and the bride walked with her own people, and they were not once seen together all the rest of the day.

One of the pleasantest excursions which we made at this time was to Landro in the Höllenstein Thal, about twelve miles from Cortina by the Austrian post-road.* On this occasion, our landlord supplied a comfortable little chaise on good springs, with a seat in front for the driver; and the chesnut appeared in smart harness, with red tassels on his head, and a necklace of little jingling bells.

With Giovanni again to drive, we started early one lovely July morning, following the course of the Upper Ampezzo valley, skirting all the length of the Tofana, and seeing its three summits in succession. Being so long in the ridge, the great height and size of this mountain can only be appreciated by those who see it from at least two sides of its vast triangle—as from the

---

* An open omnibus now leaves Cortina daily at 6·30 A.M. for Toblach, returning the same afternoon. There is also a post omnibus. (*Note to Second Edition.*)

Tre Sassi pass on the S.W., and from the high road on the East. Good walkers with time to spare may complete the tour of the mountain by ascending the Val Travernanzes, which divides the Tofana ridge from that of Monte Lagazuoi. The pyramidal peak on the side of the Tre Sassi has been repeatedly ascended by hunters from Cortina. The central peak was achieved by Dr. Grohmann in 1863; and the north peak was reached in 1867 by Mr. Bonney, who describes the view looking over in the direction of Bruneck and the Gross Venediger as one of the finest among the Eastern Alps. The highest peak, according to the latest measurements, reaches as nearly as possible to 10,724 feet.

From Cortina, the road runs for some distance at a level of about sixty feet above the bed of the Boita, and passes presently under the shadow of a kind of barber's pole painted with red and white stripes, which here juts across the road at an angle of forty-five degrees. As we prepare to drive under it, the door of a little hut adjoining, which we had taken till now for a good-sized kennel, flies suddenly open, and a small, withered, excited old man flings himself into the middle of the road, and demands forty-eight kreutzers for toll. Becoming learned in the ways of the place, we soon know that a white and red pole always stands for a toll-bar, while a black and yellow one indicates the boundary line between Austria and Italy.

From here, the road now begins to ascend and the mountains to close in; new peaks, snow streaked above and wooded below, come into view; and the great crag of Peutelstein, once crowned by a famous mediæval stronghold, shuts in the end of the valley. The old castle was levelled to the ground in 1867, and there is some talk of a modern fortress to be erected on its site. At this point, the road swings round abruptly to the right, winds up through pine-woods behind the platform on which the castle used to stand, leaves the noisy torrent far below, and, trending eastward at right angles to the Ampezzo valley, takes, in local parlance, the name of the Thal Tedesco—which, however, is not to be found in either Mayr's or Artaria's maps. Here, also, a board by the wayside informs us that we have entered the " Distretta " of Welsperg.

And now the road leads through a succession of delicious grassy glades, among pine-woods loaded with crimson and violet cones, and festooned with the weird grey-beard moss of the Upper Alps. Wild campanulas and purple gentians, deep golden Arnica blossoms, pink Daphne, and a whole world of other wild flowers, some quite new to us, here bloom in such abundance that the space of green sward on either side of the carriage-way looks as if bordered by a strip of Persian carpet.

Meanwhile, through openings in the wood, we catch occasional glimpses of great Dolomite peaks to right and left, and, emerging by and by upon an open space of meadow-land on the borders of which stands a tiny farmhouse, we see the fine pinnacles of the Cristallino (9,238 feet) rising in giant battlements beyond the sloping ground upon our right. And now the road crosses a rough torrent-bed, stony, and steep, and blinding white in the sunshine. Here we alight and

make our way across from boulder to boulder, while Giovanni leads the chesnut in and out among the shallows.

And now, as we emerge from the pine-wood, a new Dolomite—a huge, dark mournful-looking mountain ominously splashed with deep red stains—rises suddenly into towering prominence upon our left, and seems almost to overhang the road.

What mountain is this? For once, Giovanni is at fault. He thinks it must be the Croda Rossa, but he is not sure. Finding a mountain, however, here set down in Mayr's map as the Crepa Rossa, and in Artaria as the Rothwand, we are fain to conclude that it is in each case the same, with only a difference in the name.

Unlike all other Dolomites that we have yet seen, the Croda Rossa, instead of being grey and pallid, is of a gloomy brownish and purplish hue, like the mountain known as " Black Stairs," near Enniscorthy, in Ireland. Going on in the direction of Schluderbach and looking back upon the Croda Rossa, it constantly assumes a more and more threatening aspect, rising cliff above cliff towards one vast domed summit, just under which is gathered a cluster of small peaks quite steeped in blood-colour. From these, great streaks and splashes of the same hue stream down the barren precipices below, as if some great slaughter had been done there, in the old days of the world.

Passing Schluderbach, a clean-looking road-side inn, we come presently in sight of the Dürren See, a lovely little emerald green lake streaked with violet shadows and measuring about three-quarters of a mile in length.

H

Great mountains close it in on all sides, and the rich woods of the lower hills slope down to the water's edge.

THE DREI ZINNEN.

The clustered peaks, the eternal snows and glaciers of Monte Cristallo; the towering summit of the Piz Popena; and the extraordinary towers of the Drei Zinnen come one after the other into view. As for the

Drei Zinnen, they surpass in boldness and weirdness all the Dolomites of the Ampezzo. Seen through an opening between two wooded hills, they rise abruptly from behind the intervening plateau of Monte Piana, as if thrust up from the centre of the earth, like a pair of tusks. No mere description can convey to even the most apprehensive reader any correct impression of their outline, their look of intense energy, of upwardness, of bristling, irresistible force. Two barren isolated obelisks of pale, sulphurous, orange-streaked limestone, all shivered into keen scimitar-blades and shark-like teeth towards the summit, they almost defy the pencil and quite defy the pen. For the annexed illustration, however, so far as mere truthfulness of actual form goes, the writer can vouch, having sketched it very carefully from the best point along the borders of the lake.

At Landro,* a clean and comfortable inn standing alone at the head of the lake, we stayed to feed the horse and take luncheon. Here we were served with excellent cold salmon-trout from the Mesurina lake, and hot cutlets. Everything about the place looked promising. The landlord and landlady and their son, a bright lad of about seventeen, spoke only an unintelligible kind of German; but were cheerfully disposed and most obliging. Thinking that it might be a pleasant place to put up at for a few days, we enquired about rooms; but every inch of the house was occupied for the whole summer by a large party, chiefly English,

---

* Landro is the Italian name for this place, which in German is called Höhlenstein.

H 2

including a member of the Italian Club-Alpino. This gentleman, followed by a gigantic St. Bernard dog, came in while we were at luncheon, marvellously attired in a brilliant scarlet flannel blouse and high black riding boots; in which costume, followed always by his dog, he had that morning been up a difficult ice-slope of Monte Cristallo.

Luncheon over, we strolled and sketched awhile beside the fairy waters of the Dürren See—a lake into which three torrents flow, and from which no stream issues. Why it never overflows its banks, and whither the surplus water vanishes, are mysteries for which no one has yet accounted. There has been talk of hidden clefts and natural emissaries in the bed of the lake; but it is obviously unlikely, to say the least of it, that the supply and the drainage should be adjusted with such nicety. Why, therefore, the Dürren See is always full, and never too full, remains to be explained by men of science.

Of the three great mountains seen from Landro, it may be as well to mention that the Drei Zinnen * (9,833 feet) has been lately ascended by various members of the Austrian or German Alpine Clubs; that the Piz Popena (10,389 feet) was first achieved by Mr. E. R. Whitwell; and that the highest peak of Monte Cristallo (10,644 feet) was gained by Dr. Grohmann in September, 1865, from the Cristall pass, beginning on the side of the Tre Croci.

Starting from the Dürren See, the road again turns

* Of the three peaks bearing this name, only two are well seen from Landro; but as one goes up the Val d'Auronzo, all three are visible.

northward, and so runs nearly straight all the way to
Toblach, a distance of about ten more English miles.
Looking up the vista of this narrow glen from Landro,
one sees the snow-capped mountains of the Puster
Thal closing in the view.

Returning to Cortina in the pleasant afternoon, we

NEAR CORTINA.

left the carriage at a point not far from the toll-bar,
and strolled homewards by a lower path leading
through fields and meadows and past the ruins of a
curious old turreted château, one tower of which now
serves for the spire of a little church built with the
stones of the former stronghold.

Meanwhile there yet remained much to be seen and
done before we could leave Cortina. We must see the

Marmarole, hitherto completely hidden behind the Croda Malcora ; and the Mesurina Lake, famous for its otters and its salmon-trout. We must go over the Tre Croci pass, and up the Val d' Auronzo ; and above all we must visit Titian's birthplace at Pieve di Cadore. Now it seemed, so far as one could judge from maps, to be quite possible to bring all these points into a single excursion, taking each in its order, and passing a night or two on the road. In order to do this, we must follow the Ampezzo valley to Pieve di Cadore * ; then take the valley of the Piave as far as its junction with the Anziei at Tre Ponti ; thence branch off into the Val d'Auronzo ; and from Auronzo find our way back to Cortina by the Val Buona and the pass of the Tre Croci. This route, if practicable, would take us the complete circuit of the Croda Malcora, Antelao, and Marmarole, and could be done, apparently, nearly all the way by carriage road.

A consultation with old Ghedina proved that this plan was feasible as far as a place called the Casa di San Marco in the Val Buona, now accessible by means of one of the new roads in process of construction by the Italian Government. As to whether this road was or was not actually completed as far as the Casa di San Marco, he was not quite sure ; but he did not doubt that the carriage could be got along "somehow." Beyond that point, however, the new way had certainly not yet been opened, and we, as certainly, could only follow it as far as it went. He would therefore send

* There is now a post omnibus daily from Cortina to San Vito (Borca), whence the Italian diligence can be taken to Pieve di Cadore. (*Note to Second Edition.*)

saddle-horses round by the Tre Croci pass to meet us at the Casa di San Marco; the carriage coming back by way of a cart-track leading round by Landro. With these saddle-horses we could then ride up to the Mesurina Alp, and return by the Tre Croci to Cortina.

As regarded time, we could make our *giro* in either three days or two; sleeping in the one case both at Pieve di Cadore and Auronzo, or, in the other, starting early enough to spend the day at Pieve and reach Auronzo in the evening. Having heard unfavourable reports of the inn at Pieve, we decided on the latter course.*

The day we started upon this, our first long expedition, was also the day that began Giuseppe's engagement as our travelling attendant. We rose early, having ordered the carriage for seven A.M.— a roomy, well-appointed landau, drawn by a pair of capital horses, and driven by a solemn shock-headed coachman of imperturbable gravity and civility. The whole turn-out, indeed, was surprisingly good and comfortable, and would have done credit to any of the first-class hotels we had lately left behind.

The Ghedinas assembled in a body to see us off. L.'s maid, mournful enough at being left behind in a strange land, watched us from the balcony. The postmaster, the chemist, the grocer and the curé, stood together in a little knot at the corner of the piazza to see us go by.

* It would, of course, be easy to put up at Tai Cadore, where there is a perfectly unobjectionable little hostelry, about one mile from Pieve di Cadore. Persons intending to make a prolonged stay in the neighbourhood would have to do this ; we, however, not liking the idea of turning back upon our road, preferred pushing on to Auronzo.

At last, bags, rugs, and umbrellas being all in, Giuseppe jumped up to his seat on the box, the driver cracked his whip, and away we went in the midst of a chorus of " buon viaggios " from the lookers-on.

The first twelve or fourteen miles of road, as far as Tai Cadore, lay over the same ground which we had already traversed the day of our arrival at Cortina. At Tai, however, we turned aside, leaving the Monte Zucco zigzag far below, and so went up the long white road leading to the hamlet on the hill.

About halfway between the two valleys, we drew up at a little wayside church, to see a certain miracle-working crucifix said to have been found in the year 1540 in a field close by, where it was turned up accidentally by the plough. Without being (as some local antiquaries would have it believed) so ancient as either the time of the invasion of the Visigoths in A.D. 410, or that of the Huns in A.D. 432, the crucifix is undoubtedly curious, and may well have been buried for security at the time of the German invasion under Maximilian in A.D. 1508. Since that time, it is supposed to have wrought a great number of miracles ; to have sweated blood, and so stayed the pestilence of 1630 ; and in various ways to have extended an extraordinary degree of favour and protection towards the people of Cadore. The little church, originally dedicated to Saint Antonio, is now called the church of the Santissimo Crocefisso, and enjoys a high reputation throughout this part of Tyrol. The crucifix is carved in old brown wood, and the sacred image is somewhat ludicrously disfigured by a wig of real hair.

We reached Pieve di Cadore about half-past eleven A.M., delays included, and found the albergo quite as indifferent as its reputation. It was very small, very dirty, and crowded with peasants eating, drinking, and smoking. Going upstairs in search of some corner where we might leave our wraps and by-and-by take luncheon apart, we found the bedrooms so objectionable that we decided to occupy the landing. It was a comfortless place, crowded with lumber, and only a shade more airy than the rest of the house. A space was cleared, however; a couple of seats were borrowed from a neighbouring room; and the top of a great carved *cassone*, or linen-chest, was made to serve for a table. Having ordered some food to be ready by one o'clock (it being now nearly eleven) we then hastened out to see the sights of the place. The landlady's youngest daughter, an officious little girl of about twelve, volunteered as guide, and, being rejected, followed us pertinaciously from a distance.

The quaint old piazza with its gloomy arcades, its antique houses with Venetian windows, its cafés, its fountain, and its loungers, is just like the piazzas of Serravalle, Longarone, and other provincial towns of the same epoch. With its picturesque Prefettura and belfry-tower one is already familiar in the pages of Gilbert's "Cadore." There, too, is the fine old double flight of steps leading up to the principal entrance on the first floor, as in the town-hall at Heilbronn——a feature by no means Italian; and there, about midway up the shaft of the campanile, is the great, gaudy, well-remembered fresco, better meant than painted, wherein

Titian, some twelve feet in height, robed and bearded, stands out against an ultramarine background, looking very like the portrait of a caravan giant at a fair.

This picture—a gift to the Commune of Cadore from the artist who painted it — is now the only mural fresco in the town. Some years ago, one of the old houses in the piazza, now ruthlessly whitewashed, is said to have borne distinct traces of external decorations by Cesare Vecellio, the cousin and pupil of Titian.

Turning aside from the glowing piazza and following the downward slope of a hill to the left of the Prefettura, we come, at the distance of only a few yards, upon another open space, grassy and solitary, surrounded on three sides by rambling, dilapidated-looking houses, and opening on the fourth to a vista of woods and mountains.

In the midst of this little piazza stands a massive stone fountain, time-worn and water-worn, surmounted by a statue of Saint Tiziano in the robes and square cap of an ecclesiastic.* The water, trickling through two metal pipes in the pedestal beneath Saint Tiziano's feet, makes a pleasant murmuring in the old stone basin; while, half hidden behind this fountain, and leaning up as if for shelter against a larger house adjoining, stands a small whitewashed cottage upon the

---

* This picturesque little monument has now disappeared, having been superseded in 1880 by a bronze statue of heroic size designed by a Venetian artist named Del Zotto. It stands on a square pedestal, on one side of which is inscribed "A Tiziano il Cadore," and upon the other sides are enumerated the masterpieces of the great painter. (*Note to Second Edition.*)

side-wall of which an incised tablet bears the following
record :—*

NEL MCCCCLXXVII
FRA QUESTE VMILI MURA
TIZIANO VECELLIO
VENE A CELEBRE VITA
DONDE VSCIVA GIA PRESSO A CENTO ANNI
IN VENEZIA
ADDI XXVII AGOSTO
MDLXXVI.

A poor, mean-looking, low-roofed dwelling, disfigured
by external chimney-shafts and a built-out oven; lit
with tiny, blinking, mediæval windows; altogether un-
lovely; altogether unnoticeable; but—the birthplace of
Titian!

It looked different, no doubt, when he was a boy and
played outside here on the grass. It had probably a
high, steep roof, like the homesteads in his own landscape drawings; but the present old brown tiles have
been over it long enough to get mottled with yellow
lichens. One would like to know if the fountain and
the statue were there in his time; and if the water
trickled ever to the same low tune; and if the women
came there to wash their linen and fill their brazen
water jars, as they do now. This lovely green hill, at
all events, sheltered the home from the east winds; and
Monte Duranno lifted his strange crest yonder against
the southern horizon; and the woods dipped down to
the valley, then as now, where the bridle-path slopes
away to join the road to Venice.

* In the (year) MCCCCLXXVII, within these humble walls Titian Vecellio
entered (upon) a celebrated life, whence he departed, at the end of nearly a
hundred years, in Venice, on the 27th day of August, MDLXXVI.

We went up to the house, and knocked. The door was opened by a sickly, hunchbacked lad who begged us to walk in, and who seemed to be quite alone there. The house was very dark, and looked much older inside than from without. A long, low, gloomy upstairs chamber with a huge penthouse fire-place jutting into the room, was evidently as old as the days of Titian's grandfather, to whom the house originally belonged; while a very small and very dark adjoining closet, with a porthole of window sunk in a slope of massive wall, was pointed out as the room in which the great painter was born.

" But how do you know that he was born here ? " I asked.

The hunchback lifted his wasted hand with a deprecating gesture.

" They have always said so, Signora," he replied. " They have said so for more than four hundred years."

" They ? " I repeated, doubtfully.

" The Vecelli, Signora."

" I had understood that the Vecellio family was extinct."

" Scusate, Signora," said the hunchback. " The last direct descendant of ' Il Tiziano ' died not long ago —a few years before I was born; and the collateral Vecelli are citizens of Cadore to this day. If the Signora will be pleased to look for it, she will see the name of Vecellio over a shop on the right-hand side, as she returns to the Piazza."

I did look for it; and there, sure enough, over a small

shop-window I found it. It gave one an odd sort of shock, as if time were for the moment annihilated ; and I remembered how, with something of the same feeling, I once saw the name of Rubens over a shop-front in the market-place at Cologne.

I left the house less incredulous than I entered it. Of the identity of the building there has never been any kind of doubt; and I am inclined to accept with the house the identity of the room. Titian, it should be remembered, lived long enough to become, long before he died, the glory of his family. He became rich; he became noble; his fame filled Italy. Hence the room in which he was born may well have acquired, half a century before his death,—perhaps even during the lifetime of his mother—that sort of sacredness which is generally of post-mortem growth. The legend, handed down from Vecellio to Vecellio in uninterrupted succession, lays claim, therefore, to a more reliable pedigree than most traditions of a similar character

The large old house adjoining, known in Cadore as the Casa Zampieri, was the next place to be visited. It originally formed part of the Vecellio property, and it contains an early fresco, once external, but now brought inside by the enlargement of the house, and supposed to have been painted by Titian in his youth.

The hunchback offered to conduct us to this house, and, having ushered us out into the little piazza, carefully locked his own door behind him. Here, lying in wait for us, we found the officious small girl with some three or four companions of her own age, who immediately formed themselves into an uninvited body-guard, and would not be shaken off.

The hunchback rang the Zampieri bell; but no one answered. He knocked; but the echo of his knocking died away, and nothing came of it. At length he tried the door. It was only latched, and it opened instantly.

"Let us go upstairs," he said, and walked straight in.

We followed, somewhat reluctantly. The body-guard trooped in after us.

"This way," said the hunchback, already halfway up the staircase.

"But the mistress of the house," we urged, hesitatingly; "where is she?"

"Ah, *chi lo sa?* Perhaps she is out—perhaps we shall find her upstairs."

Again we followed. It was a large house, and had once upon a time been handsomely decorated. The landing was surrounded by doors and furnished with old high-backed chairs, sculptured presses, and antique oak chests big enough for two or three Ginevras to have hidden in. Our guide opened one of the doors, led us into a bare-looking kind of drawing-room, and did the honours of the place as if it all belonged to him.

"Ecco il Tiziano!" said he, pointing to a rough fresco which, though executed on the wall of the room, was set round with a common black and gold framing.

The subject, which is very simple, consists of only three figures:—a long-haired boy kneeling on one knee, and a seated Madonna, with the Child-Christ standing in her lap. These are relieved against a somewhat indefinite background of pillars and drapery. The drawing of this group is not particularly good; the colouring is thin and poor; but there is much dignity and sweetness both in the attitude and expression of the Madonna. The drapery and background have, however, suffered injury at some time or other; and, worse still, restoration. A small picture which the lad originally appeared to be presenting as a votive offering, has been altogether painted out; but its former position is clearly indicated by the attitude of the hands of the two principal figures.

According to the same respectable chain of local tradition, Titian painted this fresco at the age of eleven years. Mr. Gilbert, who knows more, and has written more, about Cadore than any of Titian's biographers, suggests that the kneeling boy is a portrait of the young painter by himself; and that he "commended himself in this manner to the Divine care" before leaving home in 1486, to become a pupil of Zuccati at Venice.

The hunchback entertained us, meanwhile with the history of the fresco; the body-guard stood gaping by; and the odious small girl amused herself by peeping into the photographic albums on the table. In the midst of it all, a door was opened at the farther end of the room, and a lady came in.

To our immense relief, she seemed to take the invasion as a matter of course, and received us as amiably as if we had presented ourselves under the properest circumstances. It may be that she is in the constant habit of finding stray foreign tourists in forcible possession of her drawing-room; but she certainly betrayed no surprise at the sight of either ourselves or our suite. She showed us some old maps and engravings of Cadore, a lithographed head of Titian, and some other worthless treasures ; and when we rose to take leave, asked for our cards.

"I value them," she said, "as souvenirs of the strangers who honour me by a visit."

The hunchback now went back to his own home, and we bent our steps towards the Duomo, always persecuted by the irrepressible little girl who, now that the hunchback had withdrawn, constituted herself our guide whether we would or no, and had it all her own way. She chattered ; she gesticulated ; she laid forcible hands upon the sketching case ; she made plunges at our parasols ; she skirmished round us, and before us, and behind us ; and kept up a breathless rush of insufferable babble.

"The Signoras were going to the Duomo ? Ecco ! They had but to follow her. *She* knew the way. *She* had known it all her life. *She* was born here ! See, that was the Prefettura. Would the Signoras like to go over the Prefettura ? Many strangers did go over the Prefettura. Yonder was the schoolhouse. *She* went to school there. She was fond of going to school. Last week she had a tooth out. It hurt dreadfully—oh ! dreadfully. It was pulled out by the medico. He lived in the piazza yonder, nearly opposite the post office. This little house here was the house of the Paroco. She had an uncle who was a Paroco ;—not here. however. At Domegge, up the valley. And she had an aunt at Cortina ; and brothers and sisters—lots of brothers and sisters, all older than herself. Her eldest sister had a baby last week—oh! such a little baby; no longer than that! Would the Signoras like to see the baby? Ah, well—here was the church. The Signoras must come in by the side door. The great door is always locked, except on Saints' days and Sundays. The side door is always open. This way—this way; and please to mind the step!"

It is a large church, quite as large as the Duomo of Serravalle, unfinished externally, bare-looking, but well-proportioned within. The chancel and transept are full of pictures, some two or three of which are reputed genuine Titians. None of these, however, though all in the style and of the school of the great master, are so strikingly fine as to declare their parentage at first sight, like the great Titian of Serravalle.

It happened, fortunately for us, that the Paroco was in the vestry. Hearing strange voices speaking a strange tongue, he came out—a handsome, gentlemanly little man of about forty-seven or fifty, with keen, well-cut features, very bright eyes, a fresh colour, and silver-grey hair. He at once entered into conversation, and was evidently well pleased to show the treasures of his church. His name and style are Don Antonio Da Via (Don being probably a corruption of Domine, a parish priest); and he has for fifteen years been paroco of this his native town. In point of taste and education he is superior to the general run of Tyrolean pastors. He takes an eager interest in all that relates to Titian and the Vecelli; and he believes Cadore to be the axis on which the world goes round.

I

The Titians in the church are two in number :—
one a large, life-size painting containing four full-
length figures ; the other an oblong, also a figure-
subject, half life size, and half length.

The first represents the Madonna and Child seated,
with S. Rocco standing on one side of the group
and S. Sebastiano on the other. S. Rocco points
as usual to the wound in his thigh. S. Sebastiano
stands in the traditional Peruginesque attitude, with
upturned face, hands bound behind his back, and his
body pierced with arrows.

The colouring has sadly faded ; the saints are not
very well-drawn ; the whole design is poor, the treatment conventional, the quality of the work early ;
and yet no student of Titian could look at it for five
minutes and doubt its authenticity. It is the figure
of the seated Madonna that stamps the work with
Titian's sign-manual. Here is the somewhat broad,
calm face, the fresh complexion, the reddish golden
hair that he delighted to paint his whole life long.
It was his favourite type of female loveliness—that
type which he developed to its ultimate perfection
in the gorgeous " Sacred and Profane Love " of the
Borghese gallery. Even the draperies of the Cadore
Madonna, although the crimson has lost its fire and
the blue has gone cold and dim, yet recall those
other glowing voluminous folds, so impossible, so
magnificent, which mark the highest ideal flight ever
yet attained in mere *pieghi*.

The present picture was doubtless executed while
Titian was yet a mere lad ; but at the same time it

bears internal evidence of having been painted after he had seen Venice and studied the works of the Venetian colourists.

Between this painting and the smaller one, there reaches a great gulf of time—a gap of perhaps fifty years. The first was the work of his boyhood; the second was the work of his age. He painted it, most likely, and presented it to the church, during one of his summer visits to his native hills. It hangs in the Vecelli chapel—a chapel dedicated to his own patron saint, S. Tiziano; and in that chapel, under that altar, it was his desire to have been finally laid to rest. He died, however, as we all know, in time of plague, at Venice; and where he died, he was, of necessity, buried.

This little picture, by which the Cadorini set unbounded store, represents Saint Tiziano and Saint Andrew adoring the Infant Christ, who lies in the lap of the Virgin. S. Tiziano, supposed to be a portrait of Titian's nephew, Marco Vecellio, kneels to the left of the spectator, in rich episcopal robes of white and gold brocade. Saint Andrew (a portrait of Titian's brother Francesco) crouches reverently on the right. Titian himself, bearing S. Tiziano's crozier, appears in attendance upon the saint, in the corner to the left; while the Virgin mother, according to popular belief, represents the wife of the painter.

The Madonna here is indifferently executed; but the Child is brought out into fine relief, and the flesh is well modelled, warm, and solid. The great feature of the picture, however, is Saint Tiziano, whose handsome, brown, uplifted face, Italian features, rich Southern complexion, and rapt, devotional expression, are in the master's purest style. The white and gold brocade of the Saint's Episcopal vestments and the subdued gold of his mitre, remind one, for their richness and solidity of texture, of the handling of Paolo Veronese. The head of Titian by himself in the left corner may be said to date the picture, and represents a man of perhaps sixty years of age. The execution of the whole is very unequal— so unequal as to suggest the idea of its having been partly executed by a scholar. In this case, however, the figures of S. Tiziano and the Infant Christ must be unhesitatingly ascribed to the hand of the master.

Besides these two pictures, the treasures of Cadore, the church contains several paintings by the brothers and nephews of Titian ; amongst others, a Last Supper by Cesare Vecellio ; a Martyrdom of St. Catherine by Orazio Vecellio ; and, foremost in merit as well as in size, four large works in tempera originally painted upon the doors of the organ by Marco Vecellio, the nephew who sat for the S. Tiziano in the altar-piece already described.

These four paintings, said the priest, had been lying for years, neglected and forgotten, in a loft to which they had been removed when taken down from the front of the organ. It had long been his desire to get them framed and hung in the church ; and now, after years of waiting, he had only just been able to carry out his design.

"A Tyrolean pastor has not many lire to spend on

the fine arts," he said smiling; "but it is done at last; and the Signoras are the first strangers who have seen them. They have not been up longer than three or four days."

These four pictures measured some sixteen feet in height by about eight in breadth, and were mounted in plain wooden frames, painted black and varnished. The outside cost of these frames, one would fancy, could scarcely have exceeded twenty lire each, or a little over three pounds English for the four. But Don Antonio had cherished his project "for years" before he was rich enough to realise it.

The temperas may be described* as four great panels, each panel decorated with a single colossal figure. Of these, Saint Matthew and Saint Mark make one pair; the Angel of the Annunciation and the Virgin, the other. With the exception of the Virgin, which is immeasurably inferior to the others, these figures are, far and away, the finest things in Cadore. For largeness of treatment, and freedom of drawing, the writer knows nothing with which to compare them, unless it be the Cartoons at South Kensington. The Angel of the Annunciation—bold, beautiful, buoyant as if just dropt down from heaven —advances on half-bended knee, with an exquisite air of mingled authority and reverence. His head and flying curls are wholly Raffaellesque. So is the grand head and upturned face of Saint Mark on one of the other panels, though sadly injured and

---

* See Crowe and Cavalcaselle's "Life of Titian," Vol. II., p. 493, where these panels are attributed to Cesare Vecellio.

obliterated. The Angel and Virgin face each other on either side of the transept, looking West; while Saint Matthew and Saint Mark occupy the same relative positions just opposite. "The Angel," said Don Antonio, "was too far separated from the Virgin; but that could not be helped, there being no other place in the church where they could be seen to so much advantage."

Having done the honours of the Sagrestia (which contained several very indifferent old pictures, including a doubtful Palma Vecchio) Don Antonio led the way up a narrow stone staircase to the Vestiario, and there, as an especial favour, permitted us to see some antique embroidered vestments and procession-banners that had been in use on great occasions from immemorial time. Much more interesting than these, however, and much more curious, was a very ancient carved and gilded Predella, or shrine, in the florid Gothic style, surmounted by a dry, Byzantine-looking Christ, and constructed with folding doors below, like a triptych. The panels of these doors were decorated outside with four small full-length paintings of the Evangelists, in a clear, brilliant, highly finished manner, the heads and general treatment recalling the style of Sandro Botticelli; while inside, the shrine contained four richly canopied niches each occupied by a small carved and painted saint, very naïve and medi-æval, like little Cimabues done in wood. This Predella belongs to a period long anterior to the Titian epoch, and adorned the high altar up to the beginning of the present century.

It was already long past the hour at which we had ordered luncheon when, having thanked Don Antonio for his courtesy, we again came out into the blinding sunshine. The insufferable little girl had now, happily, vanished; but she turned up again as soon as we re-appeared at the Albergo, buzzed about us all the time we were despatching our uncomfortable mid-day meal, and was only driven off by help of Giuseppe when we went out again presently to sketch and stroll about the town and the castle hill for another couple of hours, before pursuing our journey to Auronzo.

AURONZO AND VAL BUONA.

DOMEGGE AND LOZZO—THE LEGEND OF MONTE CORNON—TRE PONTI
—THE ANTIQUITY OF THE PIAVE—THE VAL D'AURONZO—NATIVE
POLITENESS — VILLA GRANDE AND VILLA PICCOLA — " L'ALTRO
ALBERGO"—AN UNPREPOSSESSING POPULATION—THE MARMAROLE—
A DESERTED SILVER MINE—THE NEW ROAD—DIFFICULTIES OVER-
COME—VAL BUONA—THE "CIRQUE" OF THE CRODA MALCORA—
BASTIAN THE SOLITARY—THE MESURINA ALP—A MOUNTAIN TARN
—THE TRE CROCI PASS.

CHAPTER VI.

AURONZO AND VAL BUONA.*

The view of Cadore upon which one looks back from
the bend of the road half a mile out of the town on the
way to Calalzo, and again from the Ponte della Molina,
about another mile farther on, is one of the finest of its
kind in all this part of Tyrol. At the same time, it has
in it very little of the Tyrolean element. Pictorially
speaking, it is a purely Italian subject, majestic, harmonious, classical ; with just sufficient sternness in the
mountain forms to give sublimity, but with no outlines
abrupt or fantastic enough to disturb the scenic repose
of the composition. In the foreground, we have the
ravine of the Molina spanned by a picturesque old
bridge, at the farther end of which a tiny chapel clings
to an overhanging ledge of cliff. In the middle distance, seen across an intervening chasm of misty valley,
the little far-away town of Cadore glistens on its
strange saddle-back ridge, watched over as of old by its
castle on the higher slope above. Farthest of all,

* There is now a good road from Pieve di Cadore by Domegge and the
Tre Ponti as far as Bastian's cottage, whence the traveller, following our
route, turns aside for the pass of the Tre Croci. The new road goes from
Belluno to Innichen. (*Note to Second Edition.*)

rising magnificently against the clear afternoon sky, the fine pyramidal mass of Monte Pera closes in the view. For light and shadow, for composition, for all that goes to make up a landscape in the grand style, the picture is perfect. Nothing is wanting—not even the foreground group to give it life; for here come a couple of bullock trucks across the bridge, as primitive and picturesque as if they had driven straight out of the fifteenth century. It is just such a subject as Poussin might have drawn, and Claude have coloured.

At Domegge, about three and a half miles from Cadore, we come upon a village almost wholly destroyed a few months back by fire. It is now one mass of black and shapeless ruin; but it will not long remain so, for the whole population, men, women, and little children, swarm like bees about a burnt hive, casting away rubbish, carrying loads of stones, mixing mortar, and helping to rebuild their lost homes. New foundations and new walls are already springing up, and by this present time, a second Domegge has doubtless risen on the ashes of the first.

Lozzo, the next village, about two miles farther up the valley, was burnt down in just the same way a year or two ago, and is now most unpicturesquely new, solid, and comfortable. Perhaps to be burnt out is, on the whole, the best fate that can befall the inhabitants of any of these ancient timber-built hamlets; for their dwellings are then replaced by substantial stone-built houses. As it is, what with danger from fire and danger from bergfalls, the smaller Tyrolean " paesi " are by no means safe or pleasant places to live in, and may stand comparison in point of insecurity with Portici, Torre del Greco, or any others of the Vesuvian villages.

Now the road, which has been very bad all the way from Cadore, slopes gradually down towards the bed of the torrent, passing within sight of Lorenzago to the right, and under the impending precipices of Monte Cornon to the left. Mountain and village has each its legend. Lorenzago, picturesquely perched on one of the lower slopes of Monte Cridola, claims to be the scene of the martyrdom of Saint Florian, a popular Tyrolean saint, whose intercession is supposed to be of especial efficacy in cases of fire ; while Monte Cornon is said to derive its name from an incident in the history of Cadore thus related by Mr. Gilbert :—" Along the slopes above this gorge, in the war of 1509, a division of Maximilian's troops was cautiously advancing, when the notes of a horn (corno) broke suddenly from the misty mountain side. It was but a casual herdsman sounding, as is still the custom there at certain seasons, to warn off bears ; but supposing themselves to be attacked by the Cadore people, panic seized the invaders, and they fled the way they came, over the Santa Croce pass to Sexten."—*Cadore*, p. 92.

The same rustic horn, sounded for the same purpose, may be heard here on quiet autumn evenings to this day, what time the bears come prowling down to rob orchards in the valley ; and it is remarkable that there are more bears in the district about Monte Cornon, Comelico, and the Gail Thal, than in any other part of the Alps.

A little way beyond the village of Lozzo, we cross the

Piave and continue along the left bank as far as its point of junction with the Anziei at Tre Ponti—a famous triple bridge consisting of three bold arches, each ninety feet in span, and all resting on a single central pier. To the left, winding away between richly wooded heights, lies the valley of Auronzo ; while to the right, the Upper Piave, its grey waters shrunken to half their previous volume, comes hurrying down a bare and stony channel from its source in the Carnic Alps.

And now, having tracked it for many a mile of its long course since first we saw it widening across the plain near Conegliano, we are to bid a last farewell to the Piave. It was then not very far from its grave in the Adriatic ; it is now about as distant from its cradle in the fastnesses of Monte Paralba. A curious old historical writer, one Dottore Giorgio Piloni of Belluno, who evolved a dull book in a dull style just one hundred and eighty-two years ago, speaks of the Piave not only as the largest and most important, but also as the "most ancient" river of the province, and seeks to identify it with the river Anassum* mentioned by Pliny in his chapter on the Venetian territory. He urges in proof of its antiquity, the depth of its bed and the height of its banks, "whereby," says he, "it may plainly be proved that this Piave cannot be a new river, as in other instances one sees may happen by intervention of earthquakes and other accidents." The good Doctor

---

* "Nasce la Piave nelle Alpi Taurisane sopra quel paese che per essere montuoso con greco vocabulo Cadore si chiama : si come il fiume ancora ha preso da Greci il nome di Anaxo ; che vuole in quella lingua dire fiume che per il corso suo veloce non può esser all' indietro navigato."—*Istoria di Giorgio Piloni. Libro Secondo. Venezia,* 1707.

when he wrote this had evidently never visited the scene of the great bergfall in the gorge of Serravalle, or seen the basin of the Piave at Capo di Ponte.

Taking the right bank of the Anziei, we now enter the Val d'Auronzo. The bad road which began at Cadore ends at Tre Ponti, and once more the horses have a fine, new, broad post-road beneath their feet. The sun by this time is dropping westward; the trees fling long shadows aslant the sloping sward; the gnats come out in clouds; and the air is full of evening scents and sounds. It has been a long day, and nearly twelve hours have gone by since we started from Cortina in the morning. How much longer have we yet to be upon the road before we reach Auronzo?

Being asked this question, the driver, whose politeness is such that it never permits him to give a direct answer to anything, touches his hat with his whip-handle, and replies that it is " as the Signora pleases." (Come lei piace, Signora.)

" But how many kilomètres have we yet before us?"

He coughs apologetically. " Kilomètres! Con rispetta, it is by no means a question of kilomètres. With horses like these, kilomètres go for nothing."

" Ebbene!—as a question of time, then:—how soon shall we be at Auronzo? In an hour? In an hour and a half? Before dusk?"

The driver shrugs his shoulders; looks round in a helpless way, as if seeking some means of escape; touches his hat again, and stammers:—

" Come lei piace, Signora!"

Come lei piace! It is the formula by which all his

ideas are bounded. He has no opinions of his own. He would die rather than express himself with decision about anything. Ask him what you will—the name of a village, the hour of the day, the state of the weather, his own name, age and birthplace, and he will inevitably reply: "Come lei piace." It is his invariable answer, and the effort to extract any other from him is sheer waste of breath.

The distance, however, proves to be only four miles. In about half an hour from the Tre Ponti, we come to a bend in the road, and lo! there lies a large, rambling village straggling along the near bank of the Anziei; a big mosque-like church with a glittering white dome; an older looking campanile peering above the brown roofs at the farther extremity of the place; and beyond all these, a vista of valley threaded by a deep, dark torrent fringed with sullen pine-woods. It is not the village of Auronzo, however, it is not the valley, nor the torrent, nor the pine-woods that make the beauty and wonder of the view: — it is the encircling array of mountain summits standing up rank above rank, peak beyond peak, against the clear, pale, evening sky. Farthest and strangest, at the remote end of the valley, rise the Drei Zinnen, now showing distinctly as three separate obelisks. A soft haze through which the sun is shining, hangs over the distance; and the Drei Zinnen, belted by luminous bands of filmy horizontal cloud, look like icebergs afloat in a sea of golden mist.

It is one of those rare and radiant effects that one may travel for a whole summer without seeing, and which, when they do occur, last but a few moments.

Before we had reached the first cottages, the golden light was gone, and the vapours had turned grey and ghostly.

Auronzo is divided into an upper and a lower village, known respectively as the Villa Grande and the Villa

VALLEY OF AURONZO.

Piccola. Villa Piccola, which one reaches first on entering from the Tre Ponti side, is a modern suburb to Villa Grande. The houses of this modern suburb are large and substantial, reminding one of the houses at Ober Ammergau ; and some are decorated in the same way with rough religious frescoes. To Villa Piccola belong both the large new church with the dome, and the albergo—a clean-looking house lying a little way

K

back from the road on the left hand, close against the parsonage.

Driving up to this inn, we find some four or five chaises and carettini drawn up in front of the house ; a knot of men and women gathered round the door ; faces of other men and women looking out from the upper windows ; and an unwonted air of bustle and festivity about the place. The landlady, a hard-featured dame in rusty black, standing at the door with her arms a-kimbo, shakes her head as we draw up, and does not give Giuseppe time to speak.

She cannot take us in—not she ! Couldn't take in the King of Italy, if he came this evening. Impossible. She has a wedding party from Comelico, and her house is quite full. Ecco ! There is another albergo higher up, in Villa Grande. We shall probably find room there. If not ?—well, she can't say ! She supposes we must go back the way we have come.

Giuseppe and the driver look blank. They mutter something in low voices about " l'altro albergo ; " and my ear detects an ominous emphasis on the " altro." The landlady purses up her mouth ; the travellers in possession (all in their gayest holiday clothes) survey us with an insolent air of triumph ; the coachman gathers up his reins ; and we drive on, quite discomfited.

With the scattered homesteads of Villa Piccola the good road ends abruptly, and becomes a mere stony cart-track full of ruts and rubble. Then, all at once, we find ourselves in the midst of a foul, closely-packed labyrinth of old timber houses, ruinous, smoke-blackened, dilapidated, compared with which the

meanest villages we have as yet passed through are
clean and promising. Here squalid children shout,
and sprawl, and beg ; slatternly women lean from
upper windows ; and sullen, fierce-looking men lounging in filthy doorways stare in a grim unfriendly way as
the carriage lurches past. This is Villa Grande.

Another moment, and, turning a sharp corner, we
draw up before a bare desolate-looking house standing a
little apart from the rest, with a walled-off bowling
ground on one side, in which some six or eight men are
playing at ball, and a score or two of others looking on.
This is our albergo.

We look at Giuseppe—at the house—at each other.

" Is there no other place to which we can go for the
night ? " we ask, aghast.

Giuseppe shakes his head. This and the inn at Villa
Piccola are the only two in the place.* If we do not
stay here, we have no resource but to go back to Tai
Cadore, a distance of at least fourteen, if not fifteen,
English miles.

At this crisis, out comes a tall, smiling, ungainly
woman, with an honest face and a mouth full of large,
shining teeth—an anxious, willing, cheerful body, eager
to bid us welcome ; eager to carry any number of bags
and rugs ; brimming over with good-will and civility.
She leads the way up an extremely dirty flight of stairs ;
across a still dirtier loft full of flour-sacks, cheeses, and
farming implements ; and thence up a kind of step-
ladder that leads to a landing furnished with the usual

---

* There are now three inns at Auronzo ; namely, the Albergo Alle Alpi, the
Alle Grazie, and the Vittoria. (*Note to Second Edition.*)

K 2

table and chairs, linen press and glass-cupboard. Opening off this landing are some two or three very bare but quite irreproachable bed-rooms with low whitewashed walls, and ceilings about seven feet from the ground. The floors, the bedding, the rush-bottomed chairs are all as scrupulously clean as the lower part of the establishment is unscrupulously the reverse. Carpets and curtains of course there are none. What is wanting in personal comforts is made up for, however, in the way of spiritual adornments. The walls are covered with prints of saints and martyrs in little black frames; while over the head of each bed there hangs a coloured lithograph of the Madonna displaying a plump pink heart stuck full of daggers, and looking wonderfully like a Valentine.

Here, then, we may take up our quarters and be at peace; and here, upon the landing, we are presently served with hot cutlets, coffee, eggs, and salad, all of very tolerable quality. While this meal is in preparation, we watch the players in the bowling ground. Our driver, having attended to his horses, strips off his coat and joins in the game. Giuseppe smokes his cigar, and looks gravely on. By and by, the dusk closes round; the players disperse; and we, who have to be upon the road again by 8.30 A.M., are glad to go to rest, watched over by our respective Madonnas.

Whether seen by evening grey or morning sunshine, the upper village of Auronzo is as unprepossessing, disreputable-looking a place as one would care to become acquainted with either at home or abroad. Rambling about next morning before breakfast, I saw

nothing but dirt and poverty under their least picturesque aspect. The people looked sullen, scowling,
and dissolute; the houses overcrowded; the surrounding
country not half cultivated. I afterwards learned that
the commune was poor, in debt, and over-populated;
and that the inhabitants bore an indifferent reputation.

It was pleasant enough, at all events, to drive off
again in the cool, bright morning, our horses' heads
turned once again towards the hills.

And now, Auronzo being left behind, the scenery becomes grander with each mile of the way. Every
opening gorge to right and left discloses fresh peaks
and glimpses of new horizons. The pine slopes, last
evening so gloomy, are outlined in sunshine this morning; and the torrent ripples along its bed of glittering
white pebbles, like a blue ribbon with a silver border.

The valley from this point looks like a cul de sac.
The road runs up to the foot of a great barrier of stony
débris at the base of Monte Giralba on the one side,
and there, to all appearance, ends abruptly; Monte
Rosiana (locally known as Monte Rugiana) puts forth a
gigantic buttress on the other; while the Col Agnello,
a wild pile of peaks not far short of 10,000 feet in height,
rises, an impassable barricade, between the two. It is
not till one has driven quite up to this point that the
valley, instead of being hopelessly blocked, is found to
turn off sharply to the left, narrowing to a mere gorge,
and winding round the western flank of Monte Rosiana.

Now, some little distance farther on, we pass the
desolate hamlet of Stabiziani, a cluster of half-ruined
cottages at the mouth of a wild glen leading to a

perilous and rarely-trodden pass behind the Col Agnello. And now the road plunges all at once into a dense, fragrant tract of pine-forest, musical with the singing of birds; pierced here and there by shafts of quivering sunlight; and all alive with little brown squirrels darting to and fro among the pendant fir-cones. By-and-by, a great cloven peak comes up above the tree-tops to the left, shutting out half the sunshine; and then a broad glade opens suddenly in the wood, revealing what looks at first sight like a range of new and colossal mountains, the lower spurs of which are only separated from us by the bed of the Anziei.

At this point the driver pulls up, and, turning half-round upon his box, says with the exaggerated politeness of a Master of the Ceremonies in a provincial Assembly Room :—

" Con rispetto, Signora—il Marmarole."

Being thus formally introduced to our new Dolomite, we would fain achieve a better view of it than is possible from this point. All we see of it, indeed, is a vast mass towering up indefinitely beyond the pine-forest, and, facing us, a huge slope of reddish brown earth piled to a height of some five or seven hundred feet against the mountain side. This slope of rubble, dotted over here and there with wooden sheds, marks the site of an extensive lead and silver mine, now abandoned; and a tiny hole in the face of the cliff above, no bigger apparently than a keyhole, is pointed out as the entrance to the principal shaft.

So we go on, always in the green shade of the forest, till we come to a little group of cottages known collec-

tively as the Casa di San Marco; a name recalling the
old days of Venetian sovereignty, and still marking the
frontier between Italy and Austria. Here, there being
no officials anywhere about, we pass unquestioned
under the black and yellow pole, and so arrive in a few
moments at the opening point of the new government
road which old Ghedina had given us directions to
follow as far as it went.

This new government road, carried boldly up and
through a steep hill-side of pine-forest, is considered—
and no doubt with justice—to be an excellent piece of
work; but old Holborn Hill with all the paving stones
up would have been easy driving compared with it. As
yet, indeed, it is not a road, but a rough clearing some
twenty feet in width, full of stones and rubble and slags
of knotted root, with the lately-felled pine-trunks lying
prostrate at each side, like the ranks of slain upon a
battle-field. No vehicle, it seems, has yet been brought
this way, and though we all alight instantly, it seems
doubtful whether the carriage can ever be got up. The
horses, half maddened by clouds of gad-flies, struggle
up the rugged slope, stopping every now and then to
plunge and kick furiously. The landau rocks and rolls
like a ship at sea. Every moment the road becomes
worse, and the blaze of noonday heat more intolerable.
Presently we come upon a gang of road-makers some
two hundred in number, women and children as well as
men, swarming over the banks like ants, clearing,
levelling, and stone-breaking. They pause in their
work, and stare at us as if we were creatures from
another world.

"You are the first travellers who have come up this way," says the overseer, as we pass by. "You must be Inglese!"

At length we reach a point where the road ceases altogether; its future course being marked off with stakes across a broad plateau of smooth turf. This plateau—a kind of natural arena in the midst of an upper world of pine-forest—is hemmed closely in by trees on three sides, but sinks away on the left into a wooded dell down which a clear stream leaps and sparkles. We look round, seeing no outlet, save by the way we have come, and wondering what next can be done with the carriage. To our amazement, the driver coolly takes the leader by the head and makes straight for the steep pitch dipping down to the torrent.

"You will not attempt to take the carriage down into that hole!" exclaims the writer.

"Con rispetto, Signora, there is no other way," replies the driver, deferentially.

"But the horses will break their legs, and the carriage will be dashed to pieces!"

"Come lei piace, Signora," says the driver, dimly recognising the truth of this statement.

We are standing now on the brink of the hollow, the broken bank shelving down to a depth of about thirty feet; the torrent tumbling and splashing at the bottom; and the opposite bank rising almost as abruptly beyond.

"Are we bound to get it across here?" I asked.

"Con rispetto, yes, Signora. That is to say, it can be sent back to Cortina all the way round by Auronzo and Pieve di Cadore. It is as the Signora pleases."

Now it pleases neither of the Signoras to send the carriage back by a round of something like forty-five miles ; so, after a hurried consultation, we decide to have the horses taken out, and the carriage hauled across by men. Giuseppe is thereupon despatched for a reinforcement of navvies ; and thus, by the help of some three or four stalwart fellows, the landau is lifted bodily over ; the horses are led across and re-harnessed ; and, after a little more pushing and pulling, a rough cart-track on the other side of this Rubicon is gained in safety.

Yet a few yards farther, and we emerge upon another space of grassy Alp—a green, smooth, sloping amphitheatre of perhaps some eighty acres in extent—to the East all woods ; to the West all mountains ; with one lonely little white house nestling against the verge of the forest about a quarter of a mile away. This amphitheatre is the Val Buona ; that little white house is the cottage of Bastian the wood-ranger ; yonder pale gigantic pinnacles towering in solitary splendour above the tree-tops to the rear of the cottage, are the crests of the Cristallo. But above all else, it is the view to the Westward that we have come here to see—the famous " cirque " of the Croda Malcora. And in truth, although we have already beheld much that is wild and wonderful in the world of Dolomite, we have as yet seen nothing that may compare with this.

The green sward slopes away from before our feet and vanishes in a chasm of wooded valley of unknown depth and distance ; while beyond and above this valley, reaching away far out of sight to right and left ;

piled up precipice above precipice, peak above peak; seamed with horizontal bars of snow-drift; upholding here a fold of glittering glacier; dropping there a thread of misty waterfall; cutting the sky-line with all unimaginable forms of jagged ridge and battlement, and reaching as it seems midway from earth to heaven, runs a vast unbroken chain of giant mountains. But what mountains? Familiar as we have become by this time with the Ampezzo Dolomites, there is not here one outline that either can recognise. Where, then, are we? And what should we see if we could climb yonder mighty barrier?

It takes some minutes' consideration and the help of the map, to solve these questions. Then, suddenly, all becomes clear. We are behind the Croda Malcora: directly behind Sorapis; and looking straight across in the direction of the Pelmo, which, however, is hidden by intervening mountains. The Antelao should be visible to the left, but is blocked out by the long and lofty range of the Marmarole. Somewhere away to the right, in the gap that separates this great panorama from the nearer masses of the Cristallo, lies the Tre Croci pass leading to Cortina. The main feature of the view, however, is the Croda Malcora; and we are looking at it from the back. Seen on this side, it shows as a sheer wall of impending precipice, too steep and straight to afford any resting places for the snow, save here and there upon a narrow ledge or shelf, scarce wide enough for a chamois. On the Ampezzo side, however, it flings out huge piers of rock, so that the Westward and Eastward faces of it are as unlike as though they belonged to two separate mountains. This form, as I by and by discover, is of frequent occurrence in Dolomite structure; the Civetta affording, perhaps, the most remarkable case in point.

Having looked awhile at this wonderful view, we are glad once more to escape out of the blinding sunshine into the shade of the pine-woods. Here, by the help of rugs and cloaks, we make a tent in which to rest for a couple of hours during the great heat of the day; and so, taking luncheon, studying our books and maps, listening to the bees among the wild-flowers and to the thrushes in the rustling boughs overhead, we fancy ourselves in Arcadia, or the Forest of Arden. Meanwhile, the woodman's axe is busy among the firs on the hillside, and now and then we hear the crash of a falling tree.

The forester who lives in the white cottage yonder comes by and by to pay his respects to the Signore. His name is Bastian, and he turns out to be a brother of Santo Siorpaes. He also has been a soldier, and is glad now and then, when opportunity offers, to act as guide. He lives in this lost corner of the world the whole year round. It is "molto tristo," he says; especially in winter. When autumn wanes, he provisions his little house as if for a long siege, laying in store of flour, cheese, sausage, coffee and the like. Then the snow comes, and for months no living soul ventures up from the valleys. All is white and silent, like death. The snow is as high as himself—sometimes higher; and he has to dig a trench about the house, that the light may not be blocked out of the lower windows. There was one winter, he says, not many years ago, when the falls were so sudden and so heavy, that he never went to bed at night without wondering whether he should be buried alive in his cottage before morning.

While he is yet speaking, a band of road-makers comes trooping by, whistling, and laughing, and humming scraps of songs. They are going back to work, having just eaten their mid-day mess of polenta; and their hearts are glad with wine—the rough red wine that Bastian sells at the cottage for about three kreutzers the litro, and which we at luncheon found quite undrinkable.

" The place is full of life now, at all events," says L., consolingly.

He looks after them, and shakes his head.

" Yes, Signora," he replies; " but their work here will soon be done, and then it will seem more solitary than ever."

The man is very like Santo, but has nothing of Santo's animation. The lonely life seems to have taken all that brightness out of him. His manner is sad and subdued; and when he is not speaking, he has just that sort of lost look which one sees in the faces of prisoners who have been a long time in confinement.

At two o'clock, we break up our camp and prepare to start again. The polite driver, mindful of a possible buono-mano, comes to take leave, and is succeeded by the lad Giovanni, who has journeyed up from Cortina to meet us with the promised saddle-horses. And now our old friend the tall chesnut appears upon the scene with the Pezzé side-saddle on his back, followed by an equally big black horse with the Ghedina saddle; whereupon, having Giuseppe and Giovanni in attendance, we mount and ride away—not without certain shrewd suspicions that our gallant steeds are carrying ladies for the first time. Big as they are, they climb, however, like cats, clambering in a wonderful way up the steep and stony slope of fir-forest that rises behind Bastian's cottage and leads to the Mesurina Alp beyond.

Three quarters of an hour of this rough work brings us to a higher level than we have yet reached, and lands us on an immense plateau of rich turf hemmed in on both sides by an avenue of rocky summits. Those to the right are the Cime Cadino, or Cadine-spitzen. Those on the left are the lower crags of the Cristallo mass, above which, though unseen from here, towers the gigantic Piz Popena. And this vast prairie-valley, so high, so solitary, all greenest grass below, all bluest sky above, undulating away into measureless distance, is the Mesurina Alp.* As much perhaps as a thousand head of cattle are here feeding in the rich pastures. Presently we pass the " Stabilimento," or *Vacherie* as it would be called in France;—a cluster of substantial wooden buildings, where the herdsmen live in summer, making and storing the cheeses which form so important an item in the wealth of the district.

---

* The word Alp is used here and always in its local sense, as signifying a mountain pasture. It may be as well to remark at the same time that the word "Col" stands in these parts for a hill, and is derived from Collis; while a mountain pass (called in Switzerland a Col) is here called a Forcella.

At length, when we have journeyed on and on, for what seems an interminable distance, we come upon a circular hollow in the midst of which nestles the Mesurina lake—a green transparent, tranquil tarn, fed as we are told by thirty springs, and rich in salmon trout and otters. The place is inconceivably still, beautiful, and solitary. Dark rushes fringe the borders of the lake, and are doubled by reflection. Three cows stand drowsing in the water, motionless. Not a ripple disturbs its glassy surface. Not a sound stirs the air. Yonder, where the vista opens Northwards, appear the cloudy summits of the Drei Zinnen; here, where the grassy lawn slopes down to the water's edge, the very sunshine seems asleep. The whole scene has a breathless unreality about it, as if it were a mirage, or a picture.

Having rested here awhile, we retrace our steps the whole length of the plateau, and then, dismounting, strike across on foot over a long slope of bog and rock, till we gain the mule-track leading by the Tre Croci pass to Cortina. An easy ascent winding up and round the edge of a pine forest, now carries us over the shoulder of the Cristallo, which here assumes quite a new aspect, and instead of appearing as one united mass, divides into three enormous blocks, each block in itself a mountain.

For a long way, the Eastward view still commands the range of the Marmarole and the Croda Malcora. Then by degrees, as we work round towards the West, the Marmarole is gradually lost to sight, and the Malcora crags begin to show themselves in profile. At last the summit of the pass is gained, with its three crosses; and all the familiar peaks of the Ampezzo side rise once more in magnificent array against the sunset:—to the left, the Pelmo and Rochetta; to the right, a corner of Monte Lagazuoi and the three summits of the Tofana; straight ahead the Bec di Mezzodì, Monte Nuvolau, and, beyond the gap of the Tre Sassi pass, the far-off snow slope of the Marmolata.

The road from here to Cortina, though not steep, is long and rough—so rough that we are glad to dismount presently and finish the homeward journey on foot. As we go down, a number of wayside crosses, some rudely fashioned in wood, some of rusty iron, attract our attention by their frequency on either side of the path. They are monuments to the memory of travellers lost in the sudden snow-storms which make these passes so perilous in winter-time and spring.

CAPRILE.

CHAPTER VII.

CAPRILE.

THE time at length came when we must bid goodbye to Cortina. It was a place in which many more days might have been spent with pleasure and profit. The walks were endless; the sketching was endless; the climate perfect. Still we had already overstayed the time originally set apart in our programme for the Ampezzo district; we had made all the most accessible excursions about the neighbourhood; and with the whole of that great Italian Dolomite centre that lies beyond the Tre Sassi ridge yet unexplored, it was plain that we could ill afford to linger longer on the Austrian border.

At the same time, Cortina, just because it lies upon the border, is in danger of being too hastily dismissed by travellers coming in from the Conegliano side. Marvellous as its surrounding mountains are, a stranger is apt to conclude that they but open the way to still greater marvels, and to regard the Ampezzo Thal as only the threshold of Wonderland. Even Mr. Gilbert, visiting Cortina for the first time in 1861, as he himself tells, stayed only one night there, and never ceased to regret the omission till another Tyrolean tour enabled him to repair it. For myself, looking back in memory across that intervening sea of peaks and passes which lies between Botzen and Cortina, I am inclined to place the Ampezzo Dolomites in the very first rank, both as regards position and structure. The mountains of Primiero are more extravagantly wild in outline ; the Marmolata carries more ice and snow; the Civetta is more beautiful ; the solitary giants of the Seisser Alp are more imposing ; but, taken as a group, I know nothing, whether for size, variety, or picturesqueness, to equal that great circle which, within a radius of less than twelve miles from the doors of the Aquila Nera, includes the Pelmo, Antelao, Marmarole, Croda Malcora, Cristallo and Tofana.

It was time, however, as I have said, for us to be moving onward. A practised mountaineer would doubtless find more than enough employment for a whole season within this one area ; but we, who were not mountaineers in any sense of the word, had now done our duty very fairly by the place, and so (not without reluctance) were bound to seek fresh woods and pastures new. Nothing, in short, could have been pleasanter than staying—except going.

Our next point being Caprile, it was arranged that we should ride over the Tre Sassi pass and send the luggage by caretta. Giuseppe, always economical, proposed a second caretta for the Signoras, adding that the char-road was "a little rough" on the side of Caprile. We, however, had already found it more than a little rough on the side of Cortina, and, being impressed with a lively recollection of the horrors of that drive, declined to pursue the experiment any farther.

Also, it was necessary to make sure of Ghedina's side-saddle. By taking horses and riding over the pass, we should at least get it as far as Caprile. Possession, so far, would be something gained. I am bound to confess that beyond that point our intentions, though vague, were decidedly felonious.

The morning was exquisite when we started. The caretta went first, driven by our polite friend of the other day, and we followed about half an hour later. The procession consisted of two riding-horses (Fuchs, the chesnut, and Moro, the black), a mule for the maid, the two elder Ghedinas, Giuseppe, and Giovanni. The Ghedinas were there to lead the horses when necessary, and to bring them home to-morrow; while Giovanni— inasmuch as the mule's present rider had never before mounted anything more spirited than a Sorrento donkey—had strict orders to stay by that animal's head, and never to leave his post for an instant. And indeed a less inexperienced rider might well have been excused a shade of nervousness, for the road was often steep, and often skirted the brink of very unpleasant-looking precipices; while the promised "basta," destitute alike of rail and pommel, proved to be neither more nor less than a bundle of cushions and sheepskins strapped upon a man's saddle, with no real support save a stirrup.

In this order, then, we finally started, taking our former route in the direction of Falzarego, and casting

many a backward glance at the mountains we were
leaving behind us.

Arrived once more at the little Hospice, the " Signora
Cuoca" was welcomed with acclamations. Again,
leaving the public room for the use of the men, we took
possession of the padrona's bright little kitchen ; again
the eggs and butter, the glittering brass pan, the long
brass ladle and the big apron were produced ; and again
the author covered herself with glory. It may have
been the peculiar quality of the air on this particular
pass, or it may have been the result of an exaggerated
degree of self-approbation ; but those Falzarego eggs
did certainly seem, on both occasions, to transcend in
delicacy and richness of flavour all other eggs with
which the present writer ever had the pleasure of
becoming acquainted.

It was our destiny to be overtaken by rain and mist
on the Tre Sassi. Before we left the Hospice, a few
uncertain drops were already beginning to fall, and by
the time we reached the summit, the Marmolata was
gleaming in the same ghostly way as before, through
fast-gathering vapours.

From this point, all is new. Skirting first the base
of Monte Lagazuoi, then of the abrupt crag locally
known as the Sasso d'Istria, we pass close above some
large unmelted snow-drifts, and so down into a steep
romantic glen traversed by a clear torrent " musical
with many a fall" and crossed every here and there by
a narrow bridge of roughly hewn pine trunks. Sometimes, where there is no bridge, the water sparkles all
across the path, and those on foot have to spring from

stone to stone as best they may. Dark firs and larches, growing thicker and closer as the dell dips deeper, make a green gloom overhead. Ferns, mosses, and wild flowers grow in lush luxuriance all over the steep banks, and carpet every hollow. Gaunt peaks are seen now and then through openings in the boughs, as if suspended high up in the misty air. And ever the descending path winds in and out among huge boulders covered with bushes and many-coloured lichens.

And now, as we go on, the sky darkens more and more. Then a light steady mist begins to fall; the mist turns to rain; the rain becomes a storm; and the mountains echo back a long, low peal of distant thunder. Meanwhile, the road has become very steep and slippery, and the horses keep their feet with difficulty. Then the glen turns and widens, and Castel d'Andraz—a shattered, blank-eyed ruin perched high upon a pedestal of crag—comes suddenly into sight. Steep precipices skirt the ruin on one side, and upland pastures on the other; a green valley opens away beyond; and the grassy slope beside the bridle-path is full of large wild orange lilies and crimson dog-roses that flame like jewels in a ray of sunshine which breaks at this moment through the clouds. Not even the sheets of rain still pelting pitilessly down, can blot out the wonderful beauty of the view, or reconcile me to the impossibility of stopping then and there to sketch it.

We ride on, however, for fully three-quarters of an hour more, stumbling over wet stones and sliding down steps hewn in the solid rock, till at length the little hamlet of Andraz,* half hidden among trees and precipices, and framed in overhead by a magnificent fragment of rainbow, appears in welcome proximity close beneath our feet. Another turn of the road, and we are there. The men are wet through; the horses are streaming; the rain runs in rivers off our waterproof cloaks; our umbrellas are portable gargoyles. In this state we alight at the door of Finazzer's tiny hostelry and " birraria "—a very small, clean, humble place; where, having taken off our wettest outer garments and dried ourselves thoroughly at a blazing kitchen fire, we order hot coffee and prepare to make the best of our position till the sky clears again.

Never was there such a toy parlour as that into which we are ushered on coming out of the kitchen! It is all pine-wood—new, bright, fragrant, cinnamon-coloured pine-wood, shining like gold. Walls, floor, ceiling are all alike. And it is perfectly square, too, in every way, like a beautiful little new box of Sorrento or Tunbridge ware. You might have turned it up endwise, or sidewise, or topsy-turvy, and but for the altered position of the door, I would defy the most sagacious architect to find out the difference. Then the chairs, the tables, the corner-cupboards, the clock-case, are all of the same material:—everything in the room, in short, is pine-wood, except the grate. There are certain toy-stalls in the Soho Bazaar where, at the cost of a few shillings, one may at any time buy just such wooden boxes full of just such wooden furniture, in miniature.

---

* The castle and hamlet of Andraz are also known, and frequently called by the name of Buchenstein.

By and by the rain ceases ; the clouds part ; the sun breaks out ; the horses are brought round ; and for the third time that day, we again push on for Caprile.

And now, not far below this point, the valley of Andraz debouches into the upper valley of the Cordevole —the fairest and most sylvan we have yet seen ; a valley less Italian in character than the Val d'Auronzo, more Swiss than the Ampezzo Thal ; rich in corn, maize, hemp, flax and pasture ; and bounded in the far distance by great shadowy mountains patched and streaked with snow, about whose flanks rent storm-clouds drift and gather, like the waves of an angry sea. That one of these is the Boé (which we come to know hereafter as a bastion of the Sella plateau) and that another is the Monte Padon, are facts to be taken for the present upon trust. The Marmolata is also dimly traceable now and then ; and presently a blurred, gigantic mass so enveloped in mist as to show no definite outline of any kind, is pointed out as the Civetta.

Meanwhile, the bridle-path, carried at an immense height along the shoulder of Monte Frisolet, follows every curve of the mountain—now commanding the valley of Livinallungo to the north-west—now coming in sight of a corner of the blue lake beyond Caprile to the south—now winding along the face of an almost vertical precipice—now skirting the borders of a pine-forest—now striking across a slope of greenest pasture ; and at every turn disclosing some new vista more beautiful than the last. Tiny villages, some a thousand feet below, some a thousand feet above the level of our path, are scattered far and wide, each with its little white church and picturesque campanile. Sometimes one, sometimes another of these, stands out for a few moments in brilliant sunshine ; then, as the clouds drive by, sinks away again into shadow. These vivid alternating passages of light and shade, followed by the intense gloom of another gathering storm now coming rapidly up from the valleys behind the Marmolata, altogether defy description.

And now, anxious if possible to escape another drenching, we hurry on, stared at by all who meet us, as if no such cavalcade had ever before found its way along this mountain track. Passing presently through the little village of Collaz, we attract the whole population to their doors and windows ; and two very old priests, standing by the church door, pull off their hats and bow to the ground as we ride by.

Then, as before, a light mist begins to fall, and turns presently to a heavy rainstorm which becomes heavier the longer it lasts. Then, too, the path gets steep and stony, and the horses, which have for some time been showing signs of fatigue, slip and stumble at every step. As for the black, being frightened by a flash of very vivid lightning, he becomes suddenly restive, and all but carries the writer at a single bound into the gulf below. Hereupon we dismount and, letting the horses go down by the road, make our way in rain, wind, thunder and lightning, down a narrow zigzag path at the bottom of which, some 300 feet below, appear the roofs and the church-spire of Caprile.

The Pezzés had given us up hours ago ; but seeing

our wretched little party coming along the village
street, drenched, draggled, and miserable, rush down in
a body to meet and welcome us on the threshold—old
Signora Pezzé, gentle and cordial; young Signora
Pezzé, still with a rose in her hair; the two sons whom
we already know, and all the helps and hangers-on of
the establishment. The men and horses arrive close
upon our heels; but the caretta, left behind long since
upon the road, never appears till some two hours later,
having turned quite over on the edge of a precipice and
deposited all our bags and rugs at the bottom of a steep
and muddy gulley, from which they were with difficulty
recovered.

Meanwhile a good fire is quickly lighted; wet cloth-
ing is taken to the kitchen to be dried; a hot supper is
put in preparation; and all the discomforts of the
journey are forgotten.

The Pezzé's is a large old rambling stone house,
and consists, in fact, of three houses thrown into one.*
The floors are some of stone and some of wood; the
rooms are at all sorts of levels; the windows are very
small, and full of flowers. An old metal sign—as old,
apparently, as the days of the Falieri—swings at the
corner outside; and a balcony of antique Italian
wrought iron juts out over the doorway. The public
room on the first floor is pannelled with oak and
contains a fine carved ceiling; while the landings, as
usual, are arranged as places to dine in. A set of
rooms, however, including the unwonted luxury of a

---

* This well-remembered old house is now closed, and a new hotel, the
Albergo delle Alpi, has been erected at the S. end of the village.

comfortable private sitting-room, were assigned to us on the second floor; and these we retained during all the time that we made Caprile our head-quarters. In the sitting-room we had a sofa, a round table, a cheffonier, and even a bookcase containing Guicciardini's History of Italy, and a Teatro Francese in thirty volumes. Here also were Ball's Guide-books, and Gilbert's " Dolomite Mountains " presented to Signora Pezzé by the authors. On the walls, amid a variety of little framed prints and photographs, we found portraits of F. F. T., and his sisters; in the visitors' book, the handwriting of J. A. S., of the N's, of the W's and of other friends who had passed by in foregoing years. The place, in short, was warm with pleasant memories. No wonder that it seemed like home from the first, and was home, while it lasted.

At Caprile, the traveller finds himself again in Italy. Coming down on foot in the pelting storm, we had crossed the frontier, it seemed, a little way above the zigzag. The village is but just over the border; and yet the houses and the people are as thoroughly Italian as if buried alive in the heart of the Apennines. It lies in a deep hollow at the foot of four mountains and at the junction of four valleys. The four mountains are the Monte Frisolet, the Monte Migion, the Monte Pezza, and the Monte Fernazza, locally known as the Monte Tos. The four valleys are the Val di Livinallungo, the Val Fiorentino, the Val Pettorina, and the Val d'Alleghe, or Cordevole. Each of the first three of these valleys (to say nothing of a fourth and apparently nameless tributary coming down a rocky glen behind

---

the village) brings its torrent to swell the flood of the Cordevole, which, a couple of miles lower down, flows southward through the lake of Alleghe on its way to join the Piave in the Val di Mel.

The village, murky and unprepossessing at first

VENETIAN LION AT CAPRILE.

sight, consists of one straggling street partly built upon arches. The church (which is in nowise remarkable, unless for the decorations of the organ-loft, on which is profanely painted a medallion head of the Apollo Belvedere surrounded by bouquets of flutes, fiddles and tambourines) is situate on a rising ground near the foot of the zigzag. At the farther end of the village on the side of Alleghe stands the column of St. Mark, commemorative of the old time when Caprile, like Cadore, owned the sovereignty of the Doge and the Council of Ten. The Venetian lion on the top—a battered mediæval bronze—was robbed some years ago of his wings, and the commune has talked of replacing them ever since. A carved shield on the front of the column is charged with the arms of Caprile, and beneath it a square stone tablet bears the following inscription :—

SCIPIONI . BENZONO . PAT.[S] . VEN.[S] . SER.[I] . SENATI .
VENE . COMISS . SUPER . FINIBUS . BENEFICENTISS .
CAPRILENSES . AERE . PVB. POS .   ANNO . MDCIX.

The little piazza in which this roadside monument stands is called the Contrada di San Marco. The torrent runs close behind it on the one side ; the International Dogana overlooks it on the other. In this open space, the young men of the village play at Pallo all day long. To a looker-on, this game which in summer forms the absorbing occupation of half the middle-class youth of Italy, would seem to be governed by no laws whatever, but to consist simply in tossing the ball from player to player. They use no bats ; they mark off no boundaries ; they make no running. Their interest in it, however, and their excitement, are unbounded. They begin immediately after breakfast and go on till dusk ; and when they are not playing, they are smoking cigarettes and looking on.

The Italians and Austrians profess now-a-days to be the best friends in the world, especially at these little frontier posts where they are brought into perpetual contact ; but I observed that the young men of Caprile,

although their favourite playing-ground lay just under
the windows of the Dogana, never invited the Austrian
soldiers to take part in the game. These latter, standing about with their hands in their pockets, or sitting
on the steps of the column, watched the players in a

MONTE CIVETTA.

melancholy way, and looked as if they found life dull at
Caprile.

The first sight that one goes out of doors to see is,
of course, the Civetta ; the first walk or drive one takes
is to the lake of Alleghe. As they both lie in the same
direction, and as the best view of the mountain is
gained from the road leading to the lake, if not from
the actual borders of the lake, most of the few travellers

who come this way content themselves with despatching both in a single morning, and then believe that they have " done " Caprile.

The grand façade of the Civetta—a sheer, magnificent wall of upright precipice, seamed from crown to foot with thousands of vertical fissures, and rising in a mighty arch towards the centre—faces to the north-west, looking directly up the Cordevole towards Caprile, and filling in the end of the valley as a great organ-front fills in the end of a Cathedral aisle. Towards evening, it takes all the glow of the sunset. In the morning, while the sun is yet low in the east, it shows through a veil of soft blue shade, vague and unreal as a dream. It was thus that I first saw it. I had gone rambling out through the village before breakfast, and suddenly the Civetta rose up before me like a beautiful ghost, draped in haze against a background of light. I thought it then, for simple breadth and height, for sym-metry of outline, for unity of effect, the most ideal and majestic-looking mountain I had ever seen ; and I think so still.

The lake of Alleghe lies about two miles S.S.E. of Caprile, in a green amphitheatre at the foot of the Civetta, the Monte Pezza, and the Monte Fernazza. The way to it lies along the left bank of the Cordevole, which here flows in a broad, strong current, and is bordered on the side of the char-road by a barren, pebbly tract sparsely overgrown with weeds and bushes. '

The river is dark, and deep, and brown; the lake, which is but an expansion of the river, is of a wonderful greenish blue—sapphire streaked with emerald. The

river is always rushing on at a headlong, irresistible
pace ; the lake, except when the wind sweeps straight
up the valley, is as placid as a sheet of looking-glass.
The river—an aggregate of many tributaries—is as old,
probably, as the mountains whence its many sources
flow ; the lake is new—a thing of yesterday. For a
hundred years are as yesterday in the world's history,
and where the lake of Alleghe now mirrors the clouds
and the mountains, there were orchards and cornfields,
farms and villages, only one hundred and two years ago.
A great bergfall from the Monte Pezza—or rather two
successive bergfalls—caused all this ruin, and created
all this beauty. These terrible catastrophes, as all
travellers know, are common to mountain countries ;
but among the Dolomite valleys, the bergfalls seem to
have occurred, and seem still to occur, with greater
frequency and on a more tremendous scale than elsewhere. You cannot walk or drive for ten miles in any
direction without coming upon some such scene of ruin.
It may have happened last year ; or ten, or fifty, or a
hundred years ago; or it may have happened in pre-
historic ages. Your guide in general knows nothing
about it. You ask him when it happened. He shrugs
his shoulders, and answers " Chi lo sa ?" But there,
at all events, lie the piled rocks with their buried secrets,
and often there is no outward difference to show which
fell within the memory of man, and which before the
date of man's creation.

The history of the lake of Alleghe has, however, been
handed down with unusual accuracy. The date of the
calamity, and the extent of the damage done, are re-

M

gistered in certain parish books and municipal records; and these again are supplemented by deeds and papers preserved by private families in the villages round about. Most of these families, and among them the Pezzés of Caprile, can tell of ancestors whose houses and lands were buried in the great fall of 1771.

This was how it happened :—

The Monte Pezza, of which I shall have more to say hereafter, lies to the West of the lake, being the largest of the four mountains already mentioned as surrounding the hollow in which Caprile is built. Northwards, it breaks away in abrupt precipices, culminating in a fine rocky summit some 8,000 feet above the level of the sea; but on the side nearest the lake, it slopes down in a succession of rich woods, pastures, and picturesque ravines. Skirting the opposite shore, one sees a vast, treacherous, smooth-looking slope of slatey rock, like a huge bald patch, extending all along the crest of the ridge on this side. It was from thence the fall came; it was this crest that slid away, slowly at first, and then with terrible swiftness, down into the valley.

The first disaster happened in the month of January, 1771. A charcoal-burner, it is said, who had been at work up in the woods, came down towards close of day, white and breathless, calling on those in the plain to save themselves, for the mountain was moving. A swift runner with the fear of death behind him, he fled from village to village, raising the cry as he went. But no one believed him. There were four villages then where now there is the lake. Incredulous of danger, the people of those four villages went to bed that evening as usual, and in the dead of night, the whole side of the mountain came down with a mighty rush and overwhelmed the sleepers, not one of whom escaped. Two of the villages were buried, and two were drowned; for the waters of the Cordevole, driven suddenly back, spread out as in the case of the Piave, and formed the lake as we now see it. The two buried hamlets lay close under the foot of the mountain at the Southern end of the basin, where the great masses of débris now lie piled in huge confusion. Alleghe, the chief place of the district, was situate somewhere about the middle of the lake, and is wholly lost to sight. The fourth village stood on a slope at the North end, close against that point where the Cordevole now flows into the lake.

Four more months went by; and then, on the 21st of May, there came a second downfall. This time the waters of the lake were driven up the valley with great violence, and destroyed even more property than before. In the little village which is now called Alleghe, and has been so called ever since the first Alleghe was effaced, the whole East end and choir of the present church were swept away, and the organ was carried to a considerable distance up the glen. At the same moment—for the whole lake seems to have surged up suddenly, as one wave—a tree was hurled in through the window of the room in which the curé was sitting at dinner, and the servant waiting upon him was killed on the spot. The choir has been rebuilt since then; and the organ, repaired and replaced, does duty to this day. No monument or tablet, so far as I could learn, has ever been erected to the memory of those who perished in these two great disasters; but a catafalque is dressed, and candles are lighted, and a solemn commemorative mass for the souls of the lost and dead is performed in the church at Alleghe on the 21st of May in every year.

We had been told that in winter, when the lake was frozen and the ice not too thick, and in summer on very calm days, the walls and roofs of one of the submerged villages might yet be seen, like the traditional towers of the drowned city of Lyonesse, far down below the surface of the water. An oven and a flight of stone steps, according to one of the young Pezzés, were distinctly visible; to say nothing of other less credible stories. At length, one delicious, idle, sunny afternoon, having nothing of importance to do elsewhere, we took a boat and went out upon the lake, just to test the truth of these traditions with our own eyes. Not a breath stirred when we started from Caprile; but by the time the boat was found and we were embarked in it, a light breeze had sprung up, and the whole surface of the water was in motion. Every moment, the breeze freshened and the ripple grew stronger. The withered little old woman and the rosy-cheeked girl who were rowing, bent to their oars and pulled with all their might; but, having crossed the debouchure of the river, declared themselves unable to pull us round the headland. The water by this time was quite rough, and we landed at the nearest point with difficulty. Scrambling up and along the bank for some distance, we came presently to a kind of little promontory from whence, notwithstanding the roughness of the surface, we could distinctly trace a long reach of wall and some three or four square enclosures—evidently the substructures of several houses.

"If it had been smooth enough and we could have rowed over yonder," said the old woman, pointing towards a more distant reach, "the Signoras might have seen houses with their roofs still on and their chimneys standing. They are all there—deep, deep down!"

"Have you yourself seen them?" I asked.

"Seen them? Eh, signora, I have seen them with these eyes, hundreds of times. Dio mio! there are those in Alleghe who have seen stranger sights than I. There are those living who have seen the old parish church with its belfry, all perfect, out yonder in the middle of the lake, where it is deep water. There are those living" (here her voice dropped to an awe-struck whisper) "who have heard the bells tolling under the water at midnight for the unburied dead!"

I have told the story of this little expedition out of its due place, in order to bring under one head all that I succeeded in gleaning at various times about the great bergfall of 1771. It certainly did not come off till we had been established for some two or three weeks at Caprile, and had once or twice been absent upon distant excursions.

Our first day at the Pezzé's was spent in strolling about the neighbourhood, and seeing after mules. Also in getting rid of the two Ghedinas, who were returning to Cortina with their horses, but not, if we could help it, with their side-saddle. How this delicate and difficult matter was at last negotiated matters little now. Enough that, being simple men with but few words at command, they were ultimately talked out of their convictions, and so departed—leaving the precious object behind them. We promised of course to pay for the hire of it; we promised to return it as soon as we succeeded in getting another; we promised everything possible and impossible, and were crowned with that success which is not always the reward of virtue.

"The Padre will be furious with us," said the younger brother somewhat ruefully, as he pocketed his buona-mano and turned to leave the room.

It occurred to me that this was highly probable, and that Ghedina père might not be altogether a pleasant person to deal with under those circumstances.

The poor fellows went away with evident reluctance, followed by Giovanni and the mule. We watched them down the street, and only breathed freely when they were fairly out of sight.

That same afternoon, having engaged the exclusive services of a local guide and a couple of mules for as long and as often as we might require them during our sojourn in these parts, we walked to the Col di Santa Lucia, a famous point of view in the neighbouring Val Fiorentino. Our way thither lay up yesterday's zigzag —a damp, muddy groove wriggling up the face of a steep hillside, about as pleasant to walk in as a marrow-spoon, and not much wider. Once arrived at the top, we left the valley of Andraz upon the left, and turned off towards the right—still, as yesterday, winding along the great pine-slopes of Monte Frisolet, but following the Eastward instead of the Westward face of the mountain.

It was uphill nearly all the way. Giuseppe, however, had provided two stout alpenstocks of his own cutting, and with this good help we pushed forward rapidly. The path lay half in shade and half in sunshine, commanding now a peep into the depths of the valley below; now a view of the great "slide"* on the opposite shoulder of Monte Fernazza; and now a backward glimpse of the Civetta seen above a crowd of intervening hill-tops. Thus at the end of a long pull of rather less than an hour and a half, we found ourselves some 1,500 feet above the level from which we had started, and close upon the Col di Santa Lucia—a curious saddle-backed hill like a lion couchant, keeping guard just at the curve of the Val Fiorentino. His neck is crested with a straggling line of Swiss-looking wooden houses, and his head is crowned by a picturesque little white church. He looks straight down towards the Pelmo, which closes the end of the valley magnificently, like a stupendous castle with twin-towers reaching to the clouds. One would like to know what demi-god piled those bastions; and why the lion crouched there, waiting for ever to spring upon him

* A mountain "slide" is sometimes (as in the case of the famous slide at Alpnach) a scientifically constructed incline paved with pine-trunks, down which the felled timber from the upper forests is shot into the valley without the labour and expense of transport. The slide of Monte Fernazza, however, is a mere forest-clearing about 40 feet in width and 800 or 900 in length, carried down the face of an almost perpendicular hill-side.

when he should venture out from his stronghold ; and if he is still imprisoned in the heart of the mountain. But the answer to these questions would have to be sought in the cloudland of uncreated myths.

Followed by all the children in the place, we made our way into the churchyard, and there, at the extreme end of the little promontory, sat upon the wall to enjoy the view. A glance at the map showed that the Ampezzo Thal lay just beyond the Pelmo, and that we were now looking at the mountain from exactly the reverse side. Seen from over yonder, it had resembled a mighty throne ; from here, as I have said, it showed as two enormous towers, tawny against the deep blue of the sky. A little white cloud resting lightly against the top of the farthest tower looked like a flag of truce floating from the battlements. Farther to the left, the curved beak of the Antelao, like the prow of a Roman galley, peeped out, faint and distant, above a bank of gathering cumulus. The Val Fiorentino, green and sunny and sprinkled with white villages, opened up, like a beautiful avenue, to the very foot of the Pelmo ; while northward, the valley of Codalunga met the descending slopes of Monte Gusella, and showed a streak of winding path leading down from the pass. Travellers who come that way from Cortina instead of by the Tre Sassi, have a rugged and somewhat uninteresting road to climb, and, for the sake of this one view which can afterwards be so easily reached from Caprile, lose the scenery of the exquisite upper Val Cordevole—perhaps the loveliest of all the Dolomite valleys.

Turning away at last from the view, we went in

search of the house of the Curé of Santa Lucia, upon the outer walls of which, as the story goes, there once existed a fresco by Titian—painted, it was said, in return for the hospitality with which he was entertained there when weatherbound in winter on his way to Venice. Schaubach tells how it represented " Death with his scythe, surrounded by symbols of earthly vanity "; and he furthermore adds that, having been barbarously whitewashed by some Paroco of the last century, it was with difficulty recovered. Where, however, Mr. Ball and Mr. Gilbert had, as they tell us, both failed, the present writer could scarcely hope for success. A carved " stemma," or coat of arms, over a side-door was all that the parsonage had to show, and no trace of the fresco was anywhere discernible.

I shall not soon forget that evening walk back to Caprile ; the golden splendour of the sky, the sweet scent of the new-mown hay. Neither shall I forget the two tired pedestrians, all knapsacks, beards and knicker-bockers, making for Caprile ; nor the shy little maid in the iron-spiked shoes, timid and silent, keeping goats by the pathside ; nor the goats themselves, who had no mauvaise honte, and were almost too friendly ; nor, above all, that wonderful rose-coloured vision that broke upon us as we turned down again into the valley —that vision of the Civetta, looking more than ever like a mighty organ, with its million pipes all gilded in the light of the sunset. The sky above was all light ; the wooded hills below were all shade. Monte Pezza, soaring out from a mist of purple haze, caught the rich glow upon its rocky summit. Caprile nestled snugly down in the hollow. The little village of Rocca, high on a green plateau, lifted its slender campanile against the horizon; while yet farther away, a couple of tender grey peaks, like hooded nuns, looked up to the Eastern sky, as if waiting for the evening star to rise. Then the rose-colour paled upon the lower crags; and the radiant cloud-wreath hovering midway across the face of the Civetta like an amber and golden scarf, turned grey and ghostlike. A few moments more, and the last flush faded. The sky turned a tender, greenish grey, flecked with golden films. The birds became silent in their nests. The grasshoppers burst into a shrill chorus. The torrent—steel-coloured now, with here and there a gleam of silver—rushed on, singing a wild song, and eager for the sea. Presently a feeble old peasant came across the pine-trunk bridge, staggering under a load of hay that left only his legs visible; and was followed by his wife, a brisk old woman with five hats piled upon her head, one on the top of another, and a sheaf of rakes and scythes under her arm.

So we lingered, spell-bound, till at last the gloaming came and drove us homeward. Some hours later, the clouds that we had seen gathering about the Antelao came up, bringing with them rain and heavy thunder; whereupon the ringers got up and rang the church bells all night long while the storm lasted.

AT CAPRILE.

UNSETTLED WEATHER — PROCESSIONS AND BELLS — RESOURCES OF
CAPRILE—HISTORY OF CAPRILE IN THE MIDDLE AGES—THE FREE
STATE OF ROCCA—LOCAL NOTABILITIES—THE GORGE OF SOTTOGUDA
—THE SASSO DI RONCH—CLEMENTI AND THE TWO NESSOLS—THE
GOATHERD'S CROSS—THE KING AND QUEEN OF THE DOLOMITES—A
MOUNTAIN IN RUINS—THE SASSO BIANCO—A TEMPTING PROPOSAL—
LEGENDS OF THE SASSO DI RONCH.

CHAPTER VIII.

AT CAPRILE.

THE good people of Caprile were difficult to please in the matter of weather. The bells having rung all night, the population turned out next morning in solemn procession at five A.M. to implore the Virgin's protection against storms. The clouds cleared off accordingly, and a magnificent morning followed the tempest. At midday, however, the procession formed again, and with more ceremony than before—a tall barefooted contadino in a tumbled surplice coming first, with a huge wooden crucifix; then a shabby priest with his hat on, intoning a litany; then two very small and very dirty boys in red capes, carrying unlighted lanthorns on poles; lastly, a long file of country folks marching two and two, the men first, the women next, the children last, all with their hats in their hands and all chanting. In this order they wound slowly round the village, beginning at the Contrada di San Marco.

"What is the procession for now?" I asked, turning to a respectable-looking peasant who was washing down a cart under an archway.

"They are going up to the church to pray for rain, Signora," he replied, pulling off his hat as the procession went by.

"But it rained last night," said I, "and this morning you were all praying for fine weather."

"Nay, Signora; we prayed this morning against the thunder and lightning—not against the rain," said my peasant gravely.

"Oh, I see—you want the rain, but you prefer it without the thunder."

"Yes, Signora. We want the rain badly. We have been praying against the drought these ten days past."

"But it seems to me," said I, "that you would waste less time if, instead of praying against the thunder, and the lightning, and the drought, you just asked the Madonna to put the wind round to the South-west and send forty-eight hours of steady rain immediately.

The man looked puzzled.

"It may be so, Signora," he said apologetically. "The Paroco settles all that for us—he knows best."

The poor fellow looked so humble and so serious that I turned away, quite ashamed of my own levity.

After this, we had unsettled weather for several days, during which it was invariably fine in the mornings and tempestuous towards night. This being the case, the procession came round quite regularly twice a day, to protest against the storm or the sunshine, according as the skies were foul or fair.

Meanwhile the bell-ringers must have had a hard time of it, for—much to our discomfort, though greatly to the satisfaction of the people of Caprile— the bells were going almost every night. The poor, I found, believed that this pious exercise dispersed the evil spirits of the storm; while the better sort conceived that it occasioned some kind of undulation in the air, and so broke the continuity of the electric fluid. Who would have expected to find these exploded superstitions * yet in force in any corner of Europe ? It was like being transported back into the middle ages.

To be condemned to a few days of uncertain weather at Caprile is by no means the worst fate that may befall a traveller in these parts. The place is full of delightful walks, all near enough to be enjoyed between the last shower and the next; of woods, and glens, and pastures rich in wild flowers; of easy hills for those who love climbing; of shade for the student; of trout for the angler; of ferns for the botanist. In a lovely little ravine among overhanging firs and mossy nooks of rock, not a quarter of a mile from the village, L. found specimens of

---

* "It is sayd the evil spirytes that ben in the regyon of thayre doubte moche whan they here the belles rongen: and this is the cause why the belles ben rongen whan it thondreth, and whan grete tempeste and outrages of wether happen, to the ende that the feinds and wycked spirytes sholde be abashed and flee, and cease of the movynge of tempeste."—WYNKEN DE WORDE : *The Golden Legend.*

See also the old Monkish rhyme inscribed on most mediæval bells :—

Funera plango—Fulgura frango—Sabbato pango.
Excito lentos—Dissipo ventos—Paco cruentos.

the *Cystopteris fragilis*, *Cystopteris alpina*, *Asplenium septentrionale*, and several varieties of maiden hair. And for the matter of sketching, a subject starts up before one at every turn of the path.

Nor does one, as in too many Dolomite villages and valleys, pay the penalty of starvation in exchange for all these pleasures. The food is very fairly good, and Madame Pezzé's cooking is unexceptionable. Beef, even though disguised in cinnamon, is welcome after a long and fatiguing course of veal-cutlet ; the salmon-trout of Alleghe are excellent ; the bread, the wild strawberries, the rich mountain cream are all quite delicious ; and even vegetables are not wholly unknown.

Then, besides the walks and the ferns and the sketching, Caprile—like almost every Italian place—has its special characteristics ; its local curiosities ; its own little root of mediæval history ; and these are things that do not come out unless one happens to be idling about for a few days, talking to the people, making friends with the Paroco, and borrowing all the dusty old vellum-bound books in the place. In this way, we light upon a few odd scraps of fact, more interesting to pick up, perhaps, than to relate.

Thus we learn that there were great iron-mines once at the Col di Santa Lucia ; that both Alleghe and Caprile were famous for their skilled ironsmiths and armourers ; and that they used to supply knives and swords to Venice. That exquisite old bronze door-handle wrought in the form of a mermaid, and that twisted hammer beaten out of one solid piece which I admired so much yesterday on the door of yonder dilapidated stone house at the farther end of the village, came probably from some anvil now buried at the bottom of the lake. There were forty mines once, they say, in the province of Belluno, where now only four are in operation. The old name of Caprile was Pagus Gabrielis. I could not learn that any inscriptions, urns, or mosaics had ever been found here, as at Longarone and Castel Lavazzo; so that the ancient Latin name seems to be the only Roman relic left. Towards the middle of the XV. century, the men of Caprile and Cadore united their political fortunes, and placed themselves under the protection of Venice; whereupon the Republic appointed them a governor with the title of Captain General. It was one of these Captains who erected the column of St. Mark yonder, at the extreme end of the village. At this time Caprile was a flourishing commercial centre, and the chief commune in these valleys.

Among the natural curiosities of the place, they point you out a small hole in the face of a neighbouring rock, and tell you it is the mouth of a spring inpregnated with sulphuretted hydrogen, once worked to some profit, but now abandoned. We also heard of the recent discovery of a vein of fine alabaster at a place called Digonera, a little farther up the valley, said to equal in quality the best alabaster of Tuscany.

Then, too, there is the dialect, unaccountably smacking of French in a country locked in between Venetia and Austria. Almost every little separate "paese" in these parts has its own vocabulary; and an enthusiast like Professor Max Müller might doubtless, by means

N

of a comparative analysis of these hundred-and-one dialectic varieties, extract all kinds of interesting philological flies in amber.

More curious, however, than any fact having to do with Caprile is the history of Rocca—a small village perched upon a hill just against the mouth of the Val Pettorina and fronting the precipitous northern face of Monte Pezza.* This tiny place, known in the middle ages as Rocca di Pietore, or Roccabruna, if never in the strict sense of the word a Republic, was at all events self-governed ; owing only a nominal allegiance to the Archdeacon of Capo d'Istria, and enjoying a special immunity from tax, impost, or personal service ("imposta o colta o fazioni personali"). This interesting little community, consisting of forty-five families, the men of which were nearly all armourers, was constrained in A.D. 1389 to acknowledge the suzerainty of the Visconti, who placed it under the jurisdiction of the Bellunese. Not even so, however, would Rocca resign its cherished liberties, but stipulated that all the articles of its ancient statute should be observed inviolate. The MS. original of this remarkable document, drawn out in sixty-six clauses and registered at Belluno in the year 1418, is now in the possession of Signora Pezzé of Caprile. When by and by the Visconti attempted to levy a tax upon their steel-work, the men of Rocca rebelled ; and later still, in A.D. 1659, being then subject to the Venetians and jealous as ever of their privileges, they despatched an ambassador to the Senate, reminding that august body how, " being situated on the fron-

* See Frontispiece.

tier and exposed to the attacks of enemies beyond the border, the people of Rocca had at all times testified to their patriotism with their blood, and preserved intact those privileges which were dearer to them than the pupils of their eyes." *

In all but name a republic, this little free state, smaller than either Marino or Andorre, finally lost its independence when ceded to the Austrians with the rest of Lombardo-Venetia in 1814. It now ranks as an ordinary parish in the district of Belluno; its castle has disappeared; and only four roofless walls of rough masonry in a green meadow at the foot of the hill on the side next Caprile, remain to mark the site of its former Municipal Palace. It is an ugly, gable-ended ruin, and looks like the shell of a small church.

In the way of local notabilities, Rocca has its painter, one Domenico de Biasio, whose works are supposed to have merit; while Caprile rejoices in a certain Padre Barnabas of the Capuchin order, famous for the eloquence of his sermons, which have been published in Belluno. Having neither seen the paintings nor read the sermons, I am unable to pronounce upon the excellence of either.

---

* This spirited address, a copy of which is preserved in the archives of Belluno, begins thus :—" Serenissimo Principe. La Rocca di Pietore, situata né monti più aspri e confinante con paesi esteri, in tutti i tempi esposta all' invasione dei nemici, con caratteri di Sangue ha dati segni infallibile della sua fede, e dimostrato che gli abitanti di quella, quanto più semplice e poveri di beni di fortuna, tanto più sono dotati di ardenza e prontezza a sagrificare sè stessi in servigio del principe loro ; da che è sortito che sempre dalla Serenità Vostra sono stati con clementissimo occhio riguardati conservandoli in tutti i tempi illesi ed intalli quei privilegi che gli sono più delle pupille degli occhi cari," &c., &c.

The two really remarkable natural curiosities of the place, however, are the gorge of Sottoguda and the Sasso di Ronch. Every visitor to Caprile is shown the first: we, I believe, were the first travellers who ever took the trouble to go up in search of the second.

The gorge of Sottoguda — a deep, narrow cleft between overhanging cliffs, distant about four-and-a-half, or five miles from Caprile — is in fact the upper end of the Val Pettorina, which here creeps between the lower spurs of Monte Guda and the Monte Foy. It is neither so narrow, nor so dark, nor so deep down as Pfeffers or Trient; but it reminds one of both, and, though on a smaller scale, is very fine and curious in its way. That the whole gorge is a mere crack in the rocks produced by some pre-historic natural convulsion, is evident at first sight. I even fancied that I could see how in certain places the rent cliffs might have been fitted together again, like the pieces of a child's puzzle.

The length of the gorge, which wriggles in and out like a serpent, is rather more than half-a-mile, windings included. Within this short distance, the torrent that flows through it is crossed by seventeen bridges of rough pine trunks. So abrupt are the turns and sinuosities, that never more than two of these bridges are visible at the same time, and sometimes the traveller who is only one bridge in advance is entirely lost sight of by his companions. The torrent roars along in great force, and is echoed and re-echoed in a deafening way from the cliffs on either side. The gorge is in many places not more than twelve feet wide. The precipices, at a rough guess, rise to a height of about six or seven hundred feet. The scale, after all, is not gigantic ; but the light and shadow come in grandly at certain hours, throwing one side of the defile into brilliant sunshine and the other into profoundest gloom, with an effect never to be obtained in either Pfeffers or Trient.

We first saw Sottoguda on a showery afternoon when the lights were unusually shifting and beautiful, and all the trees and bushes overhead, and all the rich red and brown and golden mosses on the rocks and boulders down below, were sparkling with rain drops. A woman standing on a slender bridge formed of a single pine-trunk thrown across a rift of rock some three hundred feet above our heads, looked down, knitting, as we wound in and out among the bridges and rapids. She smiled and spoke ; but the roar of the water was such that we could not hear her. We saw the motion of her lips, and that was all. Presently a little white goat came and thrust its head forward from behind her skirts, and also peered down upon the wayfarers below. The blue sky and the green bushes framed them round, and made a picture not soon to be forgotten.

Most travellers see Sottoguda from Caprile ; but it is approached to more advantage from the side of the Fedaya pass, and should, if possible, be first taken from that direction. Those, however, who are not equal to the fatigue of crossing the pass, may go to Sottoguda and back from Caprile in about three hours with mules, or in four hours on foot.

To go to the Sasso di Ronch, however, takes quite half a day. It is a very curious spot, and one of which the writer may claim to be in a very small way the discoverer.

Wandering about as usual before breakfast the first morning after our arrival in Caprile, and taking the road towards Alleghe, I observed a strange, solitary chimney of rock standing out against the sky, high upon the sloping shoulder of Monte Migion, about two thousand feet above the level of the valley. Seen from below, it had apparently no thickness proportionate to its height and breadth, and looked like a gigantic paper-knife stuck upright in a bed of green sward. A few trees and a couple of châlets nestled at the foot of this eccentric object; and, scaling it by these, I concluded that it could not measure less than 250 feet from base to summit. I had come out that morning to see the Civetta; but, having taken a long look at that Queen of Dolomites, I nevertheless sat down there and then upon a big boulder in a flood of burning sunshine, and, with the help of an opera-glass, sketched the Sasso di Ronch.

From that moment, I was tormented by the desire to see it more nearly. There were houses up there, so it was fair to conclude there must also be a path; and of the view it must command in at least two directions, there could be no doubt. Giuseppe, however, knew nothing about it; and none of the Pezzés had ever taken the trouble to go higher than Rocca, or Lasté, or the cross on the brink of the cliff about halfway between the two, where strangers are taken to see the view over the Civetta.

" There is nothing up yonder," said young Signora

Pezzé, contemptuously; " nothing but an old stone and a couple of poor cottages!"

But the old stone had fascinated our imaginations;

SASSO DI RONCH.

so one fine morning we sent for Clementi and the mules, and started upon our voyage of discovery.

Clementi must be introduced — Clementi and the mules. Clementi is our Caprile guide. He either belongs to the mules or the mules belong to him; it is impossible to say which. One mule is black, the other white, and both are named Nessol; which is perplexing. Fair Nessol is L.'s mule—a gentle beast, weak but willing; given to stopping and staring at the landscape in a meditative way; but liable to odd and sometimes inconvenient prejudices. Yesterday he objected to bridges, which in the gorge of Sottoguda was particularly awkward. To-day he suddenly abhors everything black, and kicks up his heels at the curé before we are out of the village. Dark Nessol, being bigger and stronger, is assigned to me. He is a self-sufficient brute; one who, in the matter of roads and turnings, invariably prefers his own opinion to that of his rider. His appetite is boundless, omnivorous, insatiable. He not only steals the young corn by the road-side and the flowers inside garden fences, but he eats poison-berries, chicken-bones, bark, egg-shells and potato-parings. He would eat the Encyclopædia Britannica, if it came in his way. L. and her mule are the best friends in the world. She feeds him perpetually with sugar, and he follows her about like a dog. My mule and I, on the contrary, never arrive at terms of intimacy. Perhaps he knows that I am the heavier weight, and resents me accordingly; perhaps he dislikes the society of ladies, and prefers carrying half-ton loads of hay and charcoal, which is the sort of thing he has been brought up to do. At all events, he refuses from the first to make himself agreeable. Both mules, however, do their work wonderfully, and climb like cats upon occasion.

Clementi is a native of Caprile, and lodges with his

old mother on the ground-floor of a big stone house in the middle of the village. He is a short, active, sturdy, black-eyed little fellow; hot tempered, ready-witted, merry, untiring, full of animation and gesture; with an honest bull-dog face, and an eye that is always laughing. He wears his trowsers tucked up round the ancles; a bunch of cock's feathers in his hat; and a bottle slung over his shoulder. It is impossible to look at him without being reminded of the clown in a Christmas pantomime. Such is Clementi; the very antipodes of Giuseppe, whom I described long since. With these two men and these two mules, we travelled henceforth as long as we remained among the Dolomites.

Setting off that bright July morning for the Sasso di Ronch, our way lies at first in the direction of Rocca. The path, however, turns aside at the ruins of the old Municipal Palace and bears away to the right, striking up at once through the fir-woods which on this side clothe the lower slopes of Monte Migion. Thus, in alternate shade and sunshine, it winds and mounts as far as the cross—a point of view on the giddy edge of an abrupt precipice facing to the South. The cliff here goes sheer down to the valley, a thousand feet or more; and Clementi tells how the cross was put there, not to mark the point of view for " Messieurs les Étrangers; " but to commemorate the death of a poor little goat-herd only eleven years of age, who, going in search of a stray kid, fell over, and was dashed to pieces before he reached the bottom.

The view from here is fine, considering at what a moderate elevation we stand. The Civetta rises before

us, grandly displayed; five valleys open away beneath
our feet; and the slated roofs of Caprile and Rocca
glisten in the morning sunshine hundreds of feet below.
A greenish-blue corner of the lake gleams just beyond
the last curve of the Val d'Alleghe; while between that
point and this, there extend, distance beyond distance,
the fir-woods, the pastures, and the young corn-slopes
of Monte Pezza.

From hence, a better path winds round towards the
North-east in the direction of Lasté—a small white
village on a mountain ledge high above the valley, looking straight over towards Buchenstein. From here the
grey old castle on its pedestal of crag, the green valley
of Andraz, and the mountains of the Tre Sassi Pass
are all visible; but the main feature of the view on this
side is the Pelmo—just as the main feature of the view
on the other side is the Civetta. Seen through a gap
in the mountains, it rises magnificently against the
horizon, looking more than ever like a gigantic fortress.
I have called the Civetta, Queen of the Dolomites; and
so, in like manner, I would call the Pelmo, King. The
one is all grace and symmetry; the other all massive-
ness and strength. It is possible to associate the idea
of fragility with the Civetta—it is possible to conceive
how that exquisite perpendicular screen with its
thousands of slender pilasters and pinnacles, might be
shivered by any great convulsion of Nature; but the
Pelmo looks as if rooted in the heart of " the great
globe itself," immoveable, till the day of the last
disruption.

For a distant view, this of the Pelmo from near Lasté

on Monte Migion, is the grandest with which I am acquainted.

From this point we next struck up across a green slope wooded like an English park, and so came out upon another path, steep and stony and glaring, which led to the cottages that I had seen from the valley.

A woman scouring a brass pan at the spring, and two others turning the yellow flax upon the hillside, stopped in their work to stare in speechless wonder. The children shouted and ran indoors, as if we were goblins.

We stayed a moment at the spring to fill our water flasks and let the mules drink.

" Have you never seen any ladies up here before ? " laughed Clementi.

" Never ! " said one of the women, throwing up her hands emphatically. " Never ! What have they come for ? "

We explained that our object was to see and sketch the Sasso up yonder.

" Il Sasso ! " she repeated, half incredulously ; " il Sasso ! "

She evidently thought us quite demented.

Another bit of rough path, another turn, and the great paper-knife rock, like a huge, solitary Menhir, is nodding over our heads.

It looks even bigger than I had expected—bigger and thinner ; but also more shapeless and less interesting. It is a marvel that the first high wind should not blow it down instantly ; but then it had this effect from below.

"Don't you think we have taken a great deal of trouble for nothing?" says L., in a tone of disappointment.

I would not acknowledge it for worlds, but I have been thinking so myself for some minutes. I push on, however, turn another corner, and arriving at the top of the Col, come suddenly upon a most unexpected and fantastic scene—a scene as of a mountain in ruins.

For not only is the whole appearance of the Sasso changed in the strangest way by being seen in profile, but behind the ridge on which the Sasso stands there is revealed a vast circular amphitheatre, like the crater of an extinct volcano, strewn with rent crags, precipices riven from top to bottom, and enormous fragments of rock, many of which are at least as big as the clock-tower at Westminster. All these are piled one upon another in the wildest confusion; all are prostrate, save one gigantic needle which stands upright in the midst of the circle, like an iceberg turned to stone.

What was the nature of this great catastrophe, and when did it happen? It could not have been a berg-fall; for the mountain slopes above are all grassy Alp, and the very summit of Monte Migion is a space of level pasture. It could not have been an eruption; for these fragments are pure Dolomite limestone, and Dolomite, it is now agreed, is not volcanic. Unable even to form a guess as to the cause of this great ruin, I can only say that, to my unscientific eyes, it looks exactly as if a volcano had burst up beneath a Dolomite summit and blown it into a thousand fragments, like a mine.

THE SASSO DI RONCH.   [P. 221.

Meanwhile here, on the ridge, apart and alone, like a solitary remnant of outer battlement left standing beside a razed fortress, rises to a height of at least 250 feet above the grass at its base, the Sasso di Ronch. Seen thus in profile, it is difficult to believe that it is the same Sasso di Ronch which one has been looking at from below. It looks like a mere aiguille, or spire, disproportionately slender for its height, and curved at the top, as if just ready to pitch over. Someone has compared the Matterhorn to the head and neck of a war-horse rearing up behind the valley of Zermatt; so might the Sasso di Ronch from this point be compared to the head and neck of a giraffe. Standing upon its knife-edge of ridge—all precipice below, all sky above, the horizon one long sweep of jagged peaks—it makes as wild and weird a subject as ever I sat down to sketch before or since!

Thus the morning passes. At noon, we rest in the shade of the Sasso to eat our frugal luncheon of bread and hard-boiled eggs; then, being refreshed, pack up the sketching traps and prepare to go home. It is not long, however, before we call another halt—this time in the midst of a beautiful open glade a little way below the cottages. Here—framed in by a foreground of velvet turf, a châlet and a group of larches, and only divided from us by the misty abyss of the Val Pettorina—rise the vertical cliffs and craggy summits of Monte Pezza. It is a ready-made sketch, and must be seized on the spot.

" There ought to be a fine view from that point yonder," I remark, mixing a pale little pool of cobalt, like a solution of turquoises, and addressing myself to no one in particular.

Hereupon Clementi, àpropos, as it would seem, of nothing, says briskly :—

" Would the Signoras like to make a first ascent ? "

" A first ascent," I repeat vaguely, adding a soften-ing drop of brown madder, and so turning the whole pool into a tender pearly grey. " What do you mean ? "

" I mean, would the Signoras like to be the first to mount to the top of the Sasso Bianco ? "

" The Sasso Bianco ! " says L., beginning to be interested in the conversation. " Where is the Sasso Bianco ? "

Clementi points to my sketch, and then to the moun-tain opposite.

" But that is the Monte Pezza ! " I exclaim.

" Scusate, Signora—the Sasso Bianco is the summit of the Monte Pezza. No traveller has ever been up there. It is new—new—new ! "

" How can it be new ? " I ask, incredulously. " It is not a very high mountain."

" Scusate ancora, Signora—it is not a mountain of the first class ; but it is high, very high, for a mountain of the second class. It is higher than either the Friso-let, the Fernazza, or the Migion."

" Still it is much less difficult than the Civetta, and the Civetta has been ascended several times. How then should the Sasso Bianco have escaped till now ? "

Because, Signora, the Sasso Bianco is too difficult for ordinary travellers, and not difficult enough for

the Club Alpino," replies Clementi, oracularly. " Il
Ball, il Tuckett, il Whitwell care nothing for a mountain which they can swallow at one mouthful."

This sounds logical. I begin to look at my mountain with more respect, and to take extra pains with

SASSO BIANCO.

my sketch. At the same time, I venture to remind
Clementi that L. and I are only " ordinary travellers "
and, as such, might find the Sasso Bianco too tough
to be swallowed in even many mouthfuls. But he
will not listen to this view of the question for a moment. If we choose to do it, we have but to say
so. He will undertake that the Signoras shall go up
" pulito."

The sketch being by this time finished, we go down, talking always of the Sasso Bianco. Clementi is eager for us to achieve the honour of a " prima ascenzione," and advocates it with all his eloquence. Giuseppe, anxious that we should attempt nothing in excess of our strength, listens gravely; puts in a question here and there; and reserves his opinion. According to Clementi, nothing can be finer than the view or easier than the ascent; but then he admits that he himself has never been higher than the upper pastures, and has never seen the view he praises so highly. Still he has gone far enough to survey the ground; he knows that we can certainly ride as far as the last group of châlets; and he is confident that the walk to the summit cannot be difficult.

On the whole, the thing sounds tempting. Our plans, however, are already laid out for a long excursion to be begun, weather permitting, to-morrow. So the subject of the Sasso Bianco, having been discussed, is for the present dismissed. Dismissed, but not forgotten. Those words " prima ascenzione " are Cabalistic, and haunt the memory strangely. They invest the Monte Pezza with a special and peculiar interest; so that it is no longer as other mountains are, but seems henceforth to have a halo round its summit.

But I must not forget the old peasant whom we met a little way below the goatherd's cross, as we went down that afternoon. He was a fine old man, still handsome, dressed in a new suit of homespun frieze and evidently well-to-do. He was sitting by the pathside. A basket and a long stick lay beside him. As we drew near, he rose and bowed; so being on foot (the men and mules following at a distance) we stopped to speak to him. He, of course, immediately asked where we had been, and where we were going. These are the invariable questions. I said that we had been up to the Sasso di Ronch.

"To the Sasso!" he repeated. "Ah, you have been up to the Sasso! Did you see the ruins of the Castle?"

I replied that, not knowing there were ruins, we had looked for nothing of the kind.

"Aye," he said, shaking his head, "and unless you knew where to find them, you would never notice them. But they are there. I have seen them myself many a time, when I was younger and could climb like you."

"Do you know to whom the Castle belonged?"

"Si, si, si—lo penso bene! Will the Signoras be pleased to sit, while I tell them all about it?"

With this he resumed his seat on the grassy bank, wiped his brow with his handkerchief, and talked away with the air of one who was accustomed to be listened to.

The Castle, he said, was built by the Visconti—the cruel Visconti of Milan. They erected it towards the close of the Fourteenth century, to overawe the "Republica" of Rocca, over which they then exercised a nominal sovereignty. But when the rule of the Visconti came to an end, the "brava" commune, fearing lest the nobles of Belluno should seize and occupy this stronghold to the ruin of the people, pulled it down,

o

leaving scarce one stone standing upon another. That was between four and five centuries ago. Then the nobles of Belluno, finding they could obtain no footing on the mountain, went and built the Castle of Andraz up yonder in the valley of Buchenstein, and there made themselves a terror to all the country. Had the Signoras seen the Castle of Andraz? Ah, well, that too was now a ruin—pulled down by the French in 1866, according to international treaty. As for the antico castello up by the Sasso, it was like an old tree of which the trunk was cut down, and only the roots left. Nothing remained of it but the foundations. Being built of the rock, they looked so like the rock that you might pass them a hundred times without observing them. There were not many people now living, he said, who knew where to look for them. When he was a young man, the contadini used to go up and dig there for hidden treasure; but they had always been frightened away by the demons. The ruins were full of demons underground, in the subterraneous dungeons, the entrances to which were now lost. They were wont to appear in the form of snakes, and they raised terrible storms of wind and thunder to drive away those who sought to discover the secrets of the ruin. Had he ever seen the demons himself? Why, no—he could not say that he had; for he had never cared to tempt the Devil by going to dig for treasure; but he had seen and heard the tempest raging up there about the top of the mountain, many and many a time, when it was fair weather down in the valley. And he had once known a man who went up at midnight on the eve of

Santo Giovanni, to dig in a certain spot where he had dreamed he should find buried gold. When he had dug a deep hole, Ecco! his spade struck against an earthen pot, and he thought his fortune was made; but when he took the lid off the pot, there came out only five small black snakes, no bigger than your finger. At this sight, being both alarmed and disappointed, he up with his spade and cut one of these little snakes in twain; and lo! in one instant, the hole that he had dug was full of snakes—big, black, venomous, twisted, hissing snakes, thousands and thousands of them, all pouring out upon him in a hideous throng, so that he had to fly for his life, and only escaped death by a miracle!

"But has nothing ever been found in the ruins?" I asked, when at the end of this story the old man paused to take breath.

"Nothing but rubbish, Signora," he replied. "A few small coins—a rusty casque or two—some fragments of armour—niente più!"

He would have talked on for an hour, if we could have stayed to listen to him; but we were in haste, and now wished him good-day. So he rose again, took off his hat, and in quaint, set terms wished us "good health, a pleasant journey, a safe return, and the blessing of God."

The rest of that afternoon was spent in laying out our route by the map; unpacking and selecting stores; and endeavouring to solve the oft-propounded problem of how to get the contents of a large portmanteau into a small black bag. For the days of caretti, landaus, and carriage-roads were over. Henceforth our ways would lie among mountain paths and unfrequented mule-tracks, and to-morrow we must start upon an expedition of at least ten days with only as much luggage as each could carry packed behind her own saddle. Giuseppe, it was arranged, should carry the sketching traps, and Clementi the provision basket. In this order we were to take a long round beginning with Cencenighe and Agordo, going thence to Primiero, Paneveggio and Predazzo, and coming home by Campidello and the Fedaja pass. In the meanwhile L.'s maid was to be left in charge of the rooms, and under the kindly care of the Pezzés.

TO AGORDO AND PRIMIERO.

DIFFICULTY OF GETTING UNDER WAY—FISHING FOR TIMBER—CEN-
CENIGHE—A VALLEY OF ROCKS—AGORDO AND ITS PIAZZA—THE
MINES OF THE VAL IMPERINA — THE DINNER "DOLOROUS" — A
SPLENDID STORM — VOLTAGO AND FRASSENE — AN "UNTRODDEN
PEAK"—THE GOSALDA PASS—A LAND OF FAMINE—MONTE PRABELLO
—THE CEREDA PASS—A JOURNEY WITHOUT AN END—CASTEL PIETRA
—PRIMIERO AT LAST—ANCIENT LINEAGE OF THE TYROLEAN INN-
KEEPERS.

CHAPTER IX.

TO AGORDO AND PRIMIERO.

HAVING risen literally with the dawn, we are on the road next morning before six, bound for Agordo. The Pezzés gather about the house-door to see us off. The Austrian officer who lodges over the way and soothes his Customs-laden soul by perpetually torturing a cracked zitter, leans out in his shirt sleeves from a second-floor window, to see us mount. He is already smoking his second, if not his third meerschaum ; and only pauses now and then to twirl his moustache with that air of serene contempt for everyone but himself which so eminently distinguishes him.

It takes some little time to strap on the bags, to say good-bye, and to induce dark Nessol to receive me upon any terms. He has a hypocritical way of standing quite still till the very moment Giuseppe is about to put me up, and then suddenly ducks away, to my immense discomfiture and the undisguised entertainment of the neighbourhood. When this performance has been repeated some six or seven times, he is hustled into a corner and pinned against the wall by main force, while I mount ignominiously at last by the help of a chair.

The road to Cencenighe lies by way of Alleghe, so that for the first five miles or more it is all familiar ground. The air is fresh, but the sky is already one blaze of cloudless sunlight. The Civetta rises before us in shadowy splendour. The larks are singing as I had thought they never sang anywhere save on the Campagna between Rome and Tivoli.

Between forty and fifty bronzed and bare-legged peasants are collecting floating timber this morning at the head of the lake. Some wade; some pilot rough rafts of tree-trunks loosely lashed together; some stand on the banks and draw the logs to shore by means of long boat-hooks. One active fellow sits his pine-trunk as if it were a horse, and paddles it to shore with uncommon dexterity. The whole scene is highly picturesque and amusing; and the men, with their shirt-sleeves rolled above their elbows, and their trowsers above their knees, look just like Neapolitan fishermen. Every now and then they all join in a shrill, prolonged cry, which adds greatly to the wildness of the effect.

Skirting the borders of the lake, we draw nearer every moment to the lower cliffs of the Civetta, and arrive at the scene of the great bergfall of 1771;—a wilderness of fallen rocks, like the battle-ground of the Titans. Somewhere beneath these mountains of débris lie the two buried villages. No one any longer remembers exactly where they stood, nor even which of the four* they were. That Alleghe lay near the middle of the present water, seems to be the only fact about which every one is confident. A solitary white house, half

* Alleghe; Riete; Marin; Fucine.

podere, half albergo, stands on a hill just above the point where the Cordevole, swelled by all the torrents of the Civetta, rushes out at the lower end of the lake and pours impetuously down the steep and narrow gorge leading to Cencenighe.

Here the path, after being carried for a long way high on the mountain side, gradually descends to the level of the river, crossing and re-crossing it continually by means of picturesque wooden bridges. Here, too, an adder, sunning itself on a heap of stones by the wayside, wriggles away at our approach, and is speedily killed by Clementi, who skips about and flourishes his stick like a maniac. Meanwhile, a tremendous South-West wind blows up the gorge like a hurricane, without in any way mitigating the pitiless blaze of the sun over-head, or the glare which is flung up at a white heat from the road underfoot.

At length, about 10.30 A.M. we arrive in sight of Cencenighe, a small village in the open flat just between the Val Cordevole and the Val di Canale. The Monte Pelsa, which is, in fact, a long, wild buttress of the Civetta ; the Cima di Pape, a volcanic peak 8,239 feet in height ; and the southward ridge of Monte Pezza, enclose it in a natural amphitheatre, the central area of which is all fertile meadow-land traversed by long lines of feathery poplars. Putting up here for a couple of hours at a poor little inn in the midst of the village, we are glad to take refuge from wind and sun in a stuffy upstairs room, while the men dine, and the mules feed, and where we take luncheon. One soon learns not rashly to venture on strange meats and drinks in these remote villages. Before starting in the morning, we now habitually provide ourselves with fresh bread and hard-boiled eggs ; and so, on arriving at a new place, ask only for cheese, wine, and a fresh lettuce from the garden. The cheese is not often very palatable, and we generally give the wine to the men ; but as something must be ordered and paid for, the purpose is answered. When we are unusually tired, or minded to indulge in luxuries, we light the Etna, and treat ourselves to Liebig soup, or tea.

Beyond Cencenighe, the character of the scenery changes suddenly. It is still the Val Cordevole, but is wholly unlike its former pastoral self either above or below Caprile. Barren precipices scarred by innumerable bergfalls close in the narrow way ; fallen boulders of enormous bulk lie piled everywhere in grand and terrible confusion ; while the road is again and again cut through huge barricades of solid débris. Frequent wayside crosses repeat the old tragic story of sudden death. The torrent, chafed and tormented by a thousand obstacles, rages below. Wild Dolomitic peaks start up here and there, are seen for a moment, and then vanish. A blind beggar-woman curled up with her crutches in the recess of a painted shrine by the roadside, uplifts a wailing voice at our approach. All is mournful ; all is desolate.

By and by, the gorge widens ; the great twin-towers of Monte Lucano and the splintered peaks of Monte Pizz come into sight ; and, like a rapid change of scene upon a mighty stage, a sunny Italian valley rich in vines and chesnuts and fields of Indian corn opens out before us.

From thence the road, winding now in shade, now in sunshine, traverses a country which would be as thoroughly Southern as the inland parts about Naples, were it not that the houses in every little village are decorated in the Tyrolean way with half-obliterated frescoes of Madonnas and Saints. Large rambling farm-houses built over gloomy arches peopled by pigs, poultry, and children, enliven the landscape with an air of slovenly prosperity quite Campanian. A wayside osteria hangs out the traditional withered bough, and announces in letters a foot long :—"Buon Acqua gratis, e Vendita di Buon Vino" (Good Water for nothing and Good Wine for sale). By and by, a scattered town and an important new-looking church with a dome and two small cupolas come into sight at the far end of the valley. This is Agordo, an archdeaconry, and the Capoluogo, or chief place, of the district.

Two long, last sultry miles of dusty flat, and we are there. A large albergo at the upper end of a piazza as big at least as Trafalgar Square, receives us on arrival—a pretentious, comfortless place, with an arcade and a café on the ground floor, and no end of half-furnished upper rooms. Being ushered upstairs by a languid damsel with an enormous chignon (for there seems to be neither master, mistress, nor waiter about the place) we take possession of a whole empty floor looking to the front, and ask, of course, the tired travellers' first question, "What can we have for dinner?"

The answer to this enquiry comes in the astounding form of a regular bill of fare. We can have anything from soup to ices, if we choose ; so, in an evil hour, we order a " real " dinner, to consist of several courses, including trout and a boiled chicken. We even talk vaguely of spending a day or two in Agordo, for the longer enjoyment of such luxurious quarters.

In the meantime, having rested, we stroll out to see the town.

Strange to say, there is no town ; there is only the piazza. Houses enough there may possibly be to make a town, if one could only bring them together, and arrange them within reasonable limits ; but here they show as a mere brick-and-mortar fringe, thinly furnishing three sides of a great desolate enclosure where all the children, and all the stray dogs, and all the Pallo players most do congregate. Three sides only ; for the fourth is wholly occupied by Count Manzoni's dilapidated villa, with its unpainted shutters, its curtainless windows, and its outside multitude of tenth-rate gods and goddesses, which crowd the sky line of the façade like an army of acrobats and ballet-girls in stone.

The church, a modern work in the Renaissance style designed by Segusini, stands near the hotel at the upper, or East end of the Piazza. The door being open, we lift the heavy leathern curtain, and walk in ; but it is like walking into " Chaos and old night." Every blind is down ; every avenue is closed against the already fading daylight. A Capuchin monk and some three or four women kneel here and there, more shadowy than the shadows. A lamp burns dimly before the high altar—a few tapers flare before the shrine of the Madonna—faint gleams of gilding, outlines of frescoes, of altar-pieces, of statues, are indistinctly visible. To gain any idea of the decorations, or even of the proportions of the church, is so impossible that we defer it altogether till to-morrow, and make, instead, the tour of the piazza.

Having done this, and having peeped into a very narrow, dirty back street running up the hill behind the town, we come home (home being the albergo of the " Miniere ") to dinner.

And here I should observe that the house is so called after the copper, lead, and zinc mines which form the commercial treasure of the district. These mines, lying at the mouth of the Val Imperina, about two miles from Agordo, belonged formerly to the Republic of Venice, and are now government property. Of the wealth of their resources there seems to be but one opinion ; yet the works are carried on so parsimoniously that the nett profit seldom exceeds 50,000 lire, or about £2,000 English, per annum. A quick-silver mine near Gosalda, about six miles off in another direction, worked by a private company, is reported to pay better.

Did I say that we came home to dinner? Ah, well! it was a sultry, languid evening ; there was thunder in the air ; and, happily, we were not very hungry. I will not dwell upon the melancholy details. Enough if I observe that the boiled chicken not only came to table in its head-dress of feathers like an African chief en grande tenue, but also with its internal economy quite undisturbed. The rest of the dishes were conceived and

carried out in the same spirit :—" Non ragionam di lor,"
&c., &c.   For my own part, I believe to this day that the
cook was a raving maniac.

That dream of spending a day or two at Agordo
vanished in the course of dinner.   We resolved to push
on as quickly as possible for Primiero ; and so, as soon
as the cloth was removed, sent for Giuseppe and ordered
the mules to be at the door by half-past six next
morning.

That night there came a tremendous storm ; the
heaviest we had yet had.   It began suddenly, with a peal
of thunder, just over the roof of the hotel, and then continued to lighten and thunder incessantly for more than
half an hour before any rain fell.   The lightning seemed
to run slantwise along the clouds in jagged streams, and
to end each time with a plunge straight down into the
earth.   These streams of electric fluid were in themselves blinding white, but the light they flashed over
the landscape was of a brilliant violet, as rich in colour
as a burst of Bengal light.   I never saw anything to
equal the vividness of that violet light, or the way in
which it not only stripped the darkness from the great
mountains on the opposite side of the valley, but
brought out with intense distinctness every separate
leaf upon the trees, every tile upon the farthest house-
tops, and every blade of grass in the piazza below.
These flashes, for the first ten minutes, followed each
other at intervals of not longer than fifteen seconds, and
sometimes of intervals of five ; so that it almost seemed
as if there were flashes of darkness as well as flashes of
light.   The church-bells, as usual, were rung as long as

the storm lasted; but the thunder-peals overlapped each other so continually, and were echoed and re-echoed in such a grand way from the amphitheatre of mountains round about, that one only heard them now and then for a moment. By and by—at the end of perhaps forty minutes—there came a deafening final explosion, as if a mountain had blown up; and after that, heavy rain, and only rain, till about two o'clock A.M.

At half-past six, however, when we rode out of Agordo, the weather was as brilliant as ever. Long fleets of white clouds were sailing overhead before the wind. The air had that delicious freshness which follows a thunderstorm in summer. The trees, the grass, the wild-flowers, even the mountains, looked as if their colours had just been dashed in with a wet brush, and so left for the sun to dry them.

Our way lay across the Cordevole bridge and then up a steep path, very narrow, partly paved, and shaded on both sides by barberry bushes, wild briars all in blossom, and nut trees already thick with clusters of new fruit. Monte Lucano, in form like a younger brother of the Pelmo, towered high into the morning mist on the one hand, and the wild peaks of Monte Pizz and Monte Agnara peered out fitfully now and then upon the other. Thus we reached and passed Voltago—a picturesque village surrounded by green firwoods and slopes of Indian corn. In the valley below gleamed Agordo, with its white dome; and against the Eastern horizon rose the pinky peaks of Monte Lasteie, the shadowy ridge of Monte Pramper, and the strange, solitary needle called the Gusella di Vescova, like a warning finger pointing to the sky.

Next came a cherry country, thick with orchards full of scarlet fruit—then a romantic ravine called the Val Molina—then the scattered village of Frassene, with its little church in the midst of a mountain prairie, surrounded by firwoods. Who would dream of finding a pianoforte manufactory in such a lost corner of the hills, or a maker of violins and contrabassi a little way lower down at Voltago? Yet at Frassene, one Giuseppe Dalla Lucia turns out pianos of respectable repute, and the fiddles, little and big, of Valentino Conedera of Voltago are said to be of unusual excellence.

And now, as we ride across this space of pleasant meadowland, the mists part suddenly overhead, and reveal a startling glimpse of three enormous pallid obelisks, apparently miles high against the blue. These are the peaks of the Sasso di Campo, one of the Primiero giants, as yet unascended, and estimated by Ball at something little short of 10,000 feet above the level of the sea. The mists part and close again; the peaks stand out for one moment in brilliant sunshine, and then melt like things of air! It is our first and last sight of the Sasso di Campo.

The path, always rising, now winds through a wooded district, stony but shady, the haunt of gorgeous butterflies. Higher still, it becomes a tunnel of greenery, only just wide and high enough for man and mule. The larches meet and rustle overhead; tiny falls trickle deliciously from rock to rock, and gush every now and then across the path; while the banks on each side are tapestried all over with rich mosses, wild strawberries, and pendent festoons of Osmunda, oak, and beech ferns. If the footway were not so steep and slippery, and the work so heavy for the mules, no place could be imagined more delicious on a day like this ; for it grows hotter every hour, as the sun climbs and the vapours roll away. But the pull is too long and too difficult ; and the path in many places resolves itself into a mere broken stair-case of wet rock up which the two Nessols, though riderless, clamber and struggle with the utmost diffi-culty.

At length one last great step is surmounted, and an immense park-like plateau scattered over with clumps of larches and firs, threaded by numberless tiny tor-rents, and radiant with wild-flowers, opens away for miles before our eyes, like a rolling sea of rich green sward. This is the summit of the Gosalda pass. The village of Gosalda, a rambling hamlet lying high on the mountain-side, facing Monte Pizzon, Monte Prabello and the valley of the Mis, is reached about two miles farther on. Here we put up for the regular midday rest at a very humble little albergo ; where, however, we are well content to take possession of a clean land-ing, a deal table, a couple of wooden chairs, and an open window commanding a magnificent view over the valley and the mountains beyond. We ask, as usual, for bread, cheese, and wine ; explaining that the wine is for the men, and that we require tea-cups and spoons for ourselves ; but the landlady, a stupid, civil body with a goître, shakes her head and stands bewildered.

" Tazze ? " she repeats, wonderingly. " Tazze ? "

Finding it impossible to make her understand what *tazze* are, I sketch a cup and spoon upon the white-washed wall; whereupon she triumphantly supplies us with two pudding basins and two metal gravy spoons of enormous size, so that we look like comic characters taking tea in a pantomime.

"Ecco! you carry fire about with you!" exclaims this child of nature, staring at the blazing Etna with the open-mouthed astonishment of a savage.

Not caring to enter into an explanation of the nature and uses of spirits of wine, I venture to remind her of the bread, and enquire if she has yet served the men with their wine.

She nods and then shakes her head again, with a pause between.

"Vino, si," she replies, oracularly. "Pane, no." (Wine, yes; bread, no.)

It seems only reasonable to suggest that, having no bread in the house, she should send out for some immediately. But no. She wags not her head this time, but her fore-finger—a gesture purely Italian. It is of no use to send out for bread. There is none to be had. There is none in the "paese." No one has any—no one in Gosalda. Not even the paroco. It all comes up from the valley—when they have any. It ought to come up twice a week; but the baker is not always punctual. It is now five days since he came last, and there is not a crust left in the village.

"But why do you not make your own bread up here in Gosalda?" I asked, when she came to the end of this astounding statement.

" Eh, Signora, we have no baker."

" And what do you eat when the baker does not come ? "

" Eh, Signora—we eat polenta."

Happily, we had a little bread in the luncheon basket; but less than usual, having given some to the mules after their hard scramble up the pass. We were better off, however, than Giuseppe and Clementi, who got nothing :—not even a dish of polenta. And this in a village numbering at least some four or five hundred souls.

The peasants of the mountain district between Agordo and Primiero seemed, so far as one could judge in a single day's journey, altogether poorer, dirtier, and more ignorant than elsewhere. Most of those whom we passed on the road, or saw at work in the fields, had goîtres ; and few understood anything but their own barbarous patois. Even the landlady of the Gosalda albergo, though she was no doubt superior to many of her neighbours, spoke very little intelligible Italian, and had no kind of local information to give. Being asked the name of the noble mountain that formed the main feature of the view before her windows, she replied first that it was the Monte Cereda ; then that it was the Sasso di Mis ; and finally admitted that she did not know for certain whether it had a name at all. Yet this was a question which she must have been continually called upon to answer. The mountain, however, as set down in Ball's map, proved to be the Monte Prabello, the highest point of which (called sometimes Il Pizz, and sometimes Il Pizzocco) rises, according to Mayr, to a height of 6733 feet above the sea-level.

A second pass—the Passo della Cereda—yet lies between us and Primiero. The distance is reported to be about two hours and a half from Gosalda, and a good mule-track all the way. The path begins pretty well, being steep but shady, and winding up between rocky banks, high hedges, and overarching trees. This, however, is too pleasant to last; and soon it begins to exhibit in an exaggerated degree all the worst features of the worst parts of the Gosalda pass. The Gosalda pass was steep; but the Cereda pass is infinitely steeper. The Gosalda pass was wet underfoot; but the Cereda pass is for miles neither more nor less than the bed of a small torrent. Nor are other and larger torrents wanting; for twice we have to dismount and make our way on foot from stone to stone across rushing streams some thirty feet in width.

The wonder is that anyone should be found to live in a place so difficult of access; yet we continually pass cottages, and clusters of cottages by the wayside; and the great valley down below is quite thickly populated. One woman standing at her garden gate nursing a wizened baby of about six months old, enquires eagerly where we come from, and if we do not find it a " brutto paese ? "

Being assured, however, that the Signoras consider it not " brutto " but " bellissimo," she is struck quite dumb with amazement.

" And where—oh! where are you going ? " is her

next question, asked with a frenzied kind of eagerness, as if her life depended on the answer.

I reply that we are going to Primiero, Predazzo, Vigo, and other places.

"To Primiero!" she repeats, breathlessly. "To Predazzo! Jesu Maria! What a number of bad roads you have before you!"

So saying, she leans out over the gate, and watches us with unfeigned compassion and wonder as long as we remain in sight.

Now the valley sinks lower, and the mountains rise higher with every step of the way. The road achieves an impossible degree of steepness. The mules, left to themselves, climb in the cleverest way, and act as pioneers to those on foot. At last comes a place which can no longer be described as a road but a barrier; being in truth the last rock-wall below the plateau to which we have all this time been mounting. Here even the mules have to be helped; and, partly by pushing, partly by pulling, reach the top at last.

And now another great prairie, somewhat like the Gosalda summit, only more wild and barren, opens away in the same manner and in the same direction, like the enchanted meadow in the fairy-tale that stretched on for ever and had no ending. A little lonely osteria in the midst of this wilderness is joyfully hailed by our famishing guides, who find here not only good wine but good white bread, and plenty of it.

It has to be a short rest, however, for the day is advancing and we have already been nine hours on the road, including halts.

" How long is it now to Primiero ? " asks Giuseppe, as we are moving off again.

To which the good woman replies in the self-same words as she of Gosalda :—

" Two hours and a half !"

As a rule, the finest wild flowers throughout these mountain districts love exposed situations, and flourish most luxuriantly on heights not far below the limit of vegetation. On the Cereda, instead of growing in rich confusion as at other places, they separate into distinct masses ; showing here as a hillside of fire-coloured lilies ; yonder as a pinky dell of ragged robin ; farther on still, as a long blue tract of wild vetch interspersed with slender spires of Canterbury bells. No painter would dare faithfully to represent these incredible slopes of alternate rose and gold and blue.

At last the path begins to dip, and our hopes to rise. Every moment we expect to see the opening of some green vista with Primiero at the end of it. Meeting a decently dressed peasant of the farmer class, however, and putting the same question to him in the same words as before, we are confounded to receive precisely the same answer :—

" Circa due ore e mezza, Signore." (About two hours and a half, ladies.)

Profoundly discouraged, we ride on after this in mournful silence. It is now more than three hours since we left Gosalda, and yet we seem to be as far as ever from Primiero. If we were not tired, if we were not hungry, if the mules were not beginning to stumble at every step, the thing would be almost comic ; but as it is, we go on funereally, following always the course of a small torrent, and skirting long pasture tracts dotted over with brown châlets.

By and by, having made another two or three miles of way, we come upon a gang of country folk at work in the new-mown hay. This time Giuseppe raises his voice, and shouts the stereotyped enquiry. The answer comes back with crushing distinctness :—

" About three hours."

I begin to think we are under the dominion of some dreadful spell. I have visions of jogging on for ever, like a party of Wandering Jews, till all four have become old, grey, and decrepit. Suddenly Clementi turns round with an eye bright with smothered glee, and says :—

" Don't you think, Signora, we should get there quicker if we turned back ? "

It is a small joke ; but it serves to make us merry over our misfortunes. After this, we put the same question to everyone we meet—to a group of women carrying faggots ; to an old man driving a pig ; to a plump priest riding " sonsily " on an ass, like Sancho Panza ; to a woodcutter going home with his axe over his shoulder, like a headsman out of livery. Each, of course, gives a different answer. One says two hours ; another two hours-and-a-half ; a third three hours ; and so on. And then all at once, when we are not in the least expecting it, we come upon a grand opening and see Castel Pietra on its inaccessible peak of cloven rock standing up straight before us. Another moment, and the valley opens out at an untold depth below—a glittering vision of chesnut-woods, villages, vineyards, and purple mountains about whose summits the storm-clouds are fast gathering.

"Ecco Primiero!" says Clementi, pointing to a

CASTEL PIETRA.

many-steepled town at the end of a long white road, still miles and miles away.

This Castel Pietra—the chromo-lithograph of which, as seen from the valley, is already familiar to most readers in Gilbert and Churchill's book—is the property of a certain Count Welsperg, by whose ancestors it was built in the old feudal times, and who still lives in Primiero. The solitary tooth of rock on which it stands has split from top to bottom some time within the last century; since when it is quite inaccessible. The present owner, when a young man, succeeded once, and once only, by the help of ropes, ladders, and workmen from Primiero, in climbing with some friends to the height of those deserted towers; but that was many a year ago, and since then the owls and bats have garrisoned them undisturbed. The castle stands, a lonely sentinel, at the opening of the great Dolomite Cul de Sac, known as the Val di Canali, and is a conspicuous object from the valley of Primiero.

The final dip down from the Cereda pass is achieved by means of a stony aud almost perpendicular road, compared with which the descent from the Ghemmi on the Leuk side is level and agreeable walking. Loose stones that roll from beneath the foot, and abrupt slopes of slippery rock, make it difficult for even pedestrians with alpenstocks; but it is worse still for the mules, which slide and struggle, and scramble in a pitiful way, being helped up behind by the ends of their tails, ignominiously.

At last we reach the level; hurry along the dusty road; pass through the ruinous-looking village of Tonadigo, and just as the church clocks are striking seven P.M., ride into Primiero. Here at the "Aquila Nera," kept by Signora Bonetti, we find rest, good food, a friendly welcome, and better rooms than the outside of the house, and above all, the entrance, would lead one to expect. That entrance is dreadful— a mere dark arch leading to a goat-stable; but then the kitchen and public rooms are on the first floor, and the visitors' rooms on the second; so that the house may be said only to begin one remove above the level of the street.

It is curious how soon one learns to be content with these humble Tyrolean albergos, and to regard as friends, and almost as equals, the kindly folk that keep them. Nor, indeed, without reason; for setting aside that perfume of antique Republicanism which seems yet to linger in the air of all that was once Venice, the Tyrolean innkeepers are, for the most part, people of ancient family who have owned lands and filled responsible offices in connection with their native communes ever since the middle ages. Thus we hear of a Ghedina of the Ampezzo holding an important military command at the beginning of the XVth century. The Giacomellis who now keep the "Nave d'Oro" at Predazzo were nobles some few hundred years ago. The Pezzés date back as far as Caprile has records to show, and take their name from the Monte Pezza, on the lower slopes of which they yet hold the remnant of their ancient estates. And the Cercenas of Forno di Zoldo, of whose inn I shall have more to say hereafter, are mentioned, as we find by Mr. Gilbert's book on Cadore, in documents more than five hundred years old. I do not know whether the Bonettis of Primiero claim either a long bourgeois pedigree or a past nobility; but they are particularly courteous and hospitable, and I see no reason for supposing them to be in any respect less well-born than the others.

It is only right that persons travelling, or intending to travel, in these valleys should be acquainted with the foregoing facts. And it would be well if they remembered they are not dealing here with innkeepers of the ordinary continental stamp; but with persons who are for the most part quite independent of the albergo as a source of profit, and ready to receive strangers with a friendliness that does not appear as an item in the bill. If the accommodation is primitive, it is at all events the best they have to offer; and it is immensely cheap. If the attendance is not first-rate, there is a pleasant homeliness about the domestic arrangements which more than makes up for any little shortcomings in other ways. The mother of the family generally cooks for her guests; the father looks after the stabling; the sons and daughters wait at table. All take a personal interest in one's comfort. All are anxious to oblige. To treat them with hauteur, or with suspicion, or to give unnecessary trouble, is both unjust and impolitic. I have seen old Signora Pezzé wounded almost to tears by the way in which a certain English party secured all their possessions under lock and key every time they ventured outside the doors. The same people, on going away, disputed every item of their moderate bill, as if, no matter how little they were charged, it was to be taken for granted that they were being imposed upon somehow.

The ultimate result of such conduct on the part of our dear country-people is sufficiently obvious. The old innkeeping families will ere long close their houses against us in disgust; a class of extortionate speculators, probably Swiss, will step in and occupy the ground; newer and smarter, but far less comfortable hotels will spring up like mushrooms in these quiet valleys; all direct communication between the native townsfolk and the travelling stranger will be intercepted; and the simplicity, the poetry, the homely charm of the Dolomite district will be gone for ever.

PRIMIERO TO PREDAZZO.

PRIMIERO AND ITS HISTORY — THE EARLY SILVER-WORKERS AND
THEIR OFFERING — TRANSACQUA AND ITS TITIAN — THE PRIMIERO
DOLOMITES—THE VAL DI CANALI—MONTE PAVIONE AND THE VETTE
DI FELTRE—MONTE ARZON—THE PONTE DELLO SCHIOS—A PRIMIERO
PROGRESSIONIST—THE COMING TENOR—SIGNOR SARTORIS AND THE
ART OF APICULTURE—THE UPPER VALLEY OF THE CISMONE—SAN
MARTINO DI CASTROZZA—A SCENE FOR A GHOST STORY—THE CIMON
DELLA PALA — THE COSTONZELLA PASS — THE HOSPICE OF PANE-
VEGGIO—THE VAL TRAVIGNOLO—PREDAZZO.

CHAPTER X.

THE town of Primiero lies partly in the plain, and partly climbs the hill upon which the church is built. The houses in the flat have a semi-Venetian character, like the houses at Ceneda and Longarone. The houses on the hill are of the quaintest German Gothic, and remind one of the steep-roofed, many-turreted medi-æval buildings in Albert Durer's backgrounds. This curious juxtaposition of dissimilar architectural styles is accounted for by the fact that Primiero, in itself more purely Italian than either Caprile or Agordo, was transferred to Austria and partly colonized by German operatives about the latter end of the 14th Century. The Tedeschi, drafted hither for the working of a famous silver mine, took root, acquired wealth, built the church, and left their impress on the place, just as the Romans left theirs in Gaul, and the Greeks in Sicily.

The early history of Primiero—how it became subject first to the Goths; then to the Lombards; next (A.D. 1027) to the Bishops of Trent; next again (A.D. 1300) to the Scaligeri of Verona; then (A.D. 1315) to Prince

Charles of Luxembourg; and finally to an Archduke of the house of Hapsburg—is but a repetition of the history of most places along the line of the Bellunese frontier. That the valley was at least twice or thrice invaded, and Castel Pietra as often besieged, by the Venetians is also matter of history. It does not appear, however, that Primiero ever became an actual appanage of the great Republic, although the neighbouring village of Transacqua (which is indeed almost a suburb of Primiero, and is only separated from the town by the Cismone and a meadow or two) was ceded to, and held by, Venice in undisputed right for a length of time both before and after the date when the rest of the valley passed into the strong grasp of Austria—a grasp unloosened to this day.

For Primiero—so Italian in its scenery, its climate, its language, its national type—is Austrian still. We passed the frontier somewhere about half-way between the village of Gosalda and the osteria on the Cereda pass; but there was no black and yellow pole to mark the boundary, and we re-entered the dominions of the Emperor Francis Joseph without knowing it.

So lately as the month of July, 1872, Primiero was as inaccessible for wheeled vehicles as Venice. Whatever there may be now, there was then no line of unbroken carriage road leading to or from the valley in any direction. Be your destination what it might, you could drive but a few miles this way, or a few miles that; and then must take to either the alpenstock or the saddle. In short, every avenue to the outer world was barred by a circle of passes, all of which were practicable to mules, but not one practicable throughout for even carettini. A fine military road is, however, now in course of construction between Primiero and Predazzo, so that a direct communication for vehicles will soon be established with Neumarkt on the Botzen and Brenner line. This road was already open last summer as far as the Hospice of San Martino, and was in progress for some miles farther. Perhaps by now it may reach as far as the Val Travignolo.*

Another excellent road runs southward from Primiero to Pontetto, the limit of the Austrian frontier; but there, unfortunately, it is joined on the Italian side by a steep and very rough mule-track which continues as far as Fonzaso. From Fonzaso, however, another carriage-road leads to Feltre, and at Feltre one is in the centre of network of fine highways radiating to Belluno, Treviso, Bassano and Trient.

Less than ten years ago, Primiero was even more primitive than now. The daily posts, we are told, came in and went out on mule-back. No rattle of wheels disturbed the silent streets; no wheel-tracks scarred the pavement. At night, the good townsfolk went about with little twinkling lanthorns, and hung an oil-lamp here and there outside their doors. Things are not quite so Arcadian now. The letter-bags are carried for at least a few miles down the valley in a light caretta; the

---

* This road has long been completed. It is traversed by a daily diligence which goes in 11 hours from Predazzo to Primiero, returning from Primiero at 5·30 P.M., and stopping for the night at San Martino di Castrozza. An omnibus also plies, twice daily (in 8 hours), between Primiero and Feltre; but this last runs only in summer. (*Note to Second Edition.*)

rattling of wheels has ceased to be regarded as a phenomenon ; a gasometer has been erected near (too near) the entrance to the town ; and the inhabitants are doing all they can to get a telegraphic wire in connection with Feltre.

The town is very clean, cheerful and picturesque. In the piazza on the flat, and in some of the side-streets (for there are side-streets in Primiero), one sees many large and really good houses. They call them Palazzos. Some of these are built over great cavernous arched entrances, and lighted by Venetian twin-windows with ogive arched tops and twisted pillars. Some are enriched with elegant balconies of wrought iron ; and on one door I observed an elaborate knocker and two handles in the form of half-length female figures of exquisite workmanship.

The German houses going up the hill—the foot-pavement of which, by the way, consists of squares of wood—are quite different. They have tiny windows filled with circular glass panes about three inches in diameter, and high steep roofs pierced by rows of dormers and surmounted by fantastic weather-cocks. The ancient Fürst Amt, with its quaint oriel turrets, loop-holed walls, mediæval windows, and rows of frescoed shields charged with faded armorial bearings, would be quite in its proper place if transported to Würtzburg or Ulm. This curious building, which stands at the top of the hill just over against the church, was erected by the early silver-workers, probably as a kind of fortified guard-house, and as a place of deposit for their store of precious metal.

PRIMIERO. [P. 258.

Many houses, both on the hill and down in the flat, are decorated externally with friezes and arabesques of a simple character ; while over almost every house-door is painted up this pious phrase :—" Christus Nobiscum Stat."

Our first day in Primiero befell upon a Sunday. The church-bells began ringing merrily before five A.M., and went on till ten. The streets were thronged with peasants in their holiday clothes ; and in the piazza sat a group of country-women with baskets of crimson cherries, little golden pears, and green lettuces for sale. It was a gay and animated scene. The men, with their knee-breeches, white stockings, conical felt hats, and jackets loosely thrown across one shoulder like a cloak, looked as if they had just stepped out of one of Pinelli's etchings. Some wore a crimson sash about the waist, and some a bunch of flowers and feathers in the hat. The women wore white cloths upon their heads tied corner-wise, and had the hair cut across the forehead in a Sévigné fringe. Their voices were curiously alike— soft, deep, and guttural. Looking in at the church-door while mass was being performed, I saw the whole nave as one sea of white head-dresses, and for the moment fancied myself peeping once more into the chapel of the Beguinage at Bruges.

It is a gloomy church ; externally more Tyrolean than German, with an unusually high steep roof and lofty spire ; internally, of a severe, well-proportioned, thirteenth-century Gothic. Two recessed and canopied state-pews of old carved oak stand on either side of the principal entrance, facing the East window and the altar; and the armorial bearings of the silver-workers are emblazoned again on the walls of the chancel.

Having heard much of a certain antique silver Mostranz (or portable shrine for the exhibition of the Host) made of the pure silver of the Primiero mines and presented to the church by these same silver-workers some six hundred years ago, we waited till the congregation had dispersed, and then asked to be permitted to see it. A grave and gentlemanly young priest received us in the sacristy, and the Mostranz was taken out of a greak oak press, as old apparently as the church itself. This curious historical relic, preserved uninjured throughout all the vicissitudes of the middle ages, stands about two feet high—a light Gothic spire, in form somewhat like the spire of Milan Cathedral; surmounted by a gilt cross; and wrought into a multitude of delicate little pinnacles enclosing tiny niches peopled with figures of Evangelists and Saints.

Our curiosity gratified, we thanked the young Paroco and took our leave; whereupon, drawing himself up in a stately fashion, he wished us " Viaggio sano, buon divertimento, e salute " :—a kind of limited benediction fitted for the dismissal of well-dressed heretics.

It was impossible not to be continually startled that Sunday morning by the repeated discharges of musketry and small cannon which kept waking the mountain echoes round about, especially just before and after high mass. These came from the little hamlet of Transacqua on the other side of the Cismone, where the villagers were making high festa in honour of the arrival of a new Paroco. Walking that way towards evening, we found a green triumphal arch erected at the opening of the Transacqua road on the farther end of the bridge, and another at the entrance to the village. The porch was also festooned with garlands and devices.

All was now still. The Paroco had gone to his new home, and the villagers to their cottages. We strolled into the empty church, and saw by a little written notice wafered against the door that it was dedicated to St. Mark—as might be expected in a parish that had once been a dependency of Venice.

"The Signoras have come to see our Titian," said a croaking voice at my elbow; "but it is too dark—too dark! It should be seen at midday, when the light comes in through the side-window."

I turned, and saw a shrivelled, slipshod sexton, all in black, with a big key in his hand. He had come to lock the church up, and found the forestieri inside.

Every insignificant little town, every obscure village that has ever belonged to Venice has its pretended Titian to show. Setting aside the Titians of Pieve di Cadore, which are unquestionably genuine, and one at Zoppé of which I shall have to tell by-and-by, there are dozens of others scattered through the country which it would be flattery to describe as even copies. There was one to be seen the other day, for instance, at Cencenighe; but having heard that it was more than doubtful, we preferred resting in the shelter of the albergo to toiling up to the church in the broiling sunshine.

The altar-piece at Transacqua is an ideal portrait of St. Mark, only the head and hands of which, however, are claimed as the work of Titian. It is said to have been presented to the church by one of the Doges of Venice. It looks a poor thing, seen thus in the gathering dusk; but the light is so bad that one may as well give its authenticity the benefit of the doubt.

The view from the bridge at evening, looking over towards Castel Pietra and the mountains at the head of the Primiero valley, is singularly wild and beautiful. The Cima Cimeda, bristling all over with peaks and pinnacles, like a porcupine; the Sass Maor, a mighty double-headed monster, compared by Mr. Leslie Stephen to the upraised finger and thumb of a gigantic hand; the Cima di Ball, so called after the dauntless author of the "Alpine Guide;" and a long array of other summits, many of which are nameless to this day, here climb against the sky in strangest outline, and take the last glow of the Western sun.

I name them here from after knowledge; but, so many and so bewildering are these Primiero Dolomites, that it is not till one has been a day or two in the place, and has seen them again and again from various points of view, that one comes to identify them with anything like certainty. The Sass Maor—a corruption of Sasso Maggiore, or Great Rock—must, however, be excepted from this general assertion. It is a mountain which, once seen, can never be mistaken for any other; but which at the same time, is only to be viewed under its most extraordinary aspect from either the Val Pravitale or the Val di Canali—two diverging forks of the great upper valley behind Castel Pietra.

This use of the word " Canali," as applied to streams and torrents flowing in their own natural beds, affords a curious instance, among many others, of how the impress of Venetian thought yet lingers throughout these parts of Southern Tyrol. To the citizen of Venice, every river and rivulet was a canal; and where Venice gave her laws, she gave her phraseology also. But this by the way.

We devoted the Monday following our arrival to the Val di Canali, which is undoubtedly the great sight of Primiero. The way thither lies through Tonadigo, along the road by which we came down that weary Saturday evening, and up the stony steep crowned by Castel Pietra. Once at the top, we bear away almost due north, leaving to the right the path leading to the Cereda pass, and striking up behind the castle along the left bank of a rapid torrent rushing down toward the valley.

Having followed this track for about three-quarters of an hour, we emerge upon an open space of grassy lawn about a mile in breadth by perhaps a mile and a-half in length, at the upper end of which stands a modest white house surrounded by sheds and farm-buildings. This little summer residence has been built of late years by Count Welsperg, who also owns a "palazzo" in Primiero, and whose ancestors (once seigneurs of all the valley, with power of life and death over their vassals) erected yonder castle which, perched on its inaccessible rock like St. Simeon Stylites on his solitary pillar, yet keeps watch and ward at the mouth of the valley. Dark fir-slopes enclose this pleasant prairie round about; the torrent brawls unseen in a bushy hollow to the left; cows and goats browse here and there on the green turf; while the whole pastoral scene is "set," as it were in a cirque of Dolomite peaks of the first magnitude—a cirque with which the Circa Malcora, grand as it is, will not bear a moment's comparison. For the mountains surrounding the Val Buona lie out in a wide amphitheatre; but here the shattered walls of Dolomite, all grey and sulphur-streaked, and touched with rusty red, close in upon the valley in two long serried ranks, not more than a mile and a-half apart at their widest point, and narrowing till they meet in the form of an acute angle at the head of the glen.

Here, where the sward is smooth and the space yet broad between, two converging lines of peaks are already arrayed before our eyes—one extending nearly due East and West; the other running up from the South-East to meet it. The first is far the grandest. Beginning with the Cima Cimeda—from behind which the Sass Maor shoots out its extraordinary impending thumb, more off the perpendicular than the leaning tower of Pisa—the chain leads on in one unbroken sweep, giving first a more distant glimpse of the Pala di San Martino; coming next upon the magnificent

Cima di Fradusta; next after that upon two nameless lower peaks broken up into sheafs of splintered arrowheads; lastly upon the Cima di Canali, apparently loftiest of all the range as seen from this point.*

* The loftiest of all the Primiero peaks (and indeed of all known Dolomites, except the Marmolata, which is supposed to exceed it by about 50 or 60 feet) is the Cimon della Pala, rising 11,000 feet above the level of the sea. But the Cimon della Pala is not seen from the Val di Canali, but lies up north of the Pala di San Martino in the upper valley of the Cismone. The relative heights of those peaks visible from the Val di Canali, as far as at present ascertained, are as follows :—Sass Maor about 10,000 feet ; Pala di San Martino, 10,643 feet ; Cima di Fradusta, something over 10,500 feet ; Cima di Canali, about the height of the last named, but probably a few feet loftier. The height of the Sasso di Campo, which closes the head of the valley, is estimated at about 9,900 feet. Of these, so far as I have been able to gather from Alpine-Club authorities, the Sass Maor, Pala di San Martino, and Sasso di Campo have certainly not yet been ascended. In the last published edition of Ball's Guide to the Eastern Alps, 1870, p. 454, the Val di Canali is thus described :—" The main branch of the Cismone descending from nearly due N., receives a torrent from the N.E., issuing from Val di Canali. In the fork between these two branches rise the wonderful group of Dolomite peaks which must ever make this one of the most extraordinary of mountain valleys. Whatever fantastic forms that rock may assume elsewhere, they are here surpassed in boldness and strangeness. Of the five or six highest, all much exceeding 10,000 feet in height, there is but one that seems accessible. The others are mere towers or obelisks of rock, with sheer vertical faces, or else, as the highest peak, fashioned like a ruinous wall, abruptly broken away at one end, and cleft at frequent intervals along the ridge by chasms that appear perfectly impassable. In rock-climbing it is never safe to declare any place impracticable without actual trial. Narrow ledges and clefts give footing to a bold climber on many a seemingly impracticable declivity ; but the writer's impression as to the Primiero peaks is confirmed by two of the most experienced mountaineers, Mr. F. F. Tuckett and Melchior Anderegg."

This reference to Mr. Tuckett's verdict is also alluded to in Mr. Leslie Stephen's article on "The Peaks of Primiero" in the Alpine Journal for February, 1870. The same story was repeated to myself by a Primiero guide, with the further addition that " il Tuckett " had said the Sass Maor could never be climbed "till a bridge was thrown across the chasm that divides the lower from the higher peak." All these tales, however, Mr. Tuckett, in reply to my own direct enquiry, emphatically refutes ; adding that he never

No glaciers* find a resting place among these perpendicular precipices. Only a narrow ledge outlined in white, or a tiny intermediate plateau sheeted with dazzling snow, serves here and there to mark the line of eternal frost.

Two small but very curious features in the scene deserve mention : these are two circular holes, one just piercing the top of a solitary sabre-blade splinter jutting out from a buttress of the anonymous peak next before the Cima di Canali on the left of the valley; and another precisely similar peep-hole piercing a precisely similar sabre-blade jutting out from a spur of the Sasso Ortiga on the right of the valley, precisely opposite.

critically examined the Primiero peaks except the Pala di San Martino, the Cimon della Pala, and the Cima di Fradusta, which last he ascended alone. —A. B. E.

Note to Second Edition :—The following are the latest and most accurate determinations of the Primiero peaks by the new Austrian survey :—

Cima di Vezzana	3191 mètres.
Cimon della Pala . . .	3186 ,,
Pala di San Martino . .	2997 ,,
Cima di Fradusta . . .	2930 to 2937 mètres.
Rosetta	2740 mètres.
Cima di Ball . . .	2693 ,,
Cima Cimeda . . .	2499 ,,
Sass Maor	2816 ,,
Passo della Val di Roda .	2568 ,,
Figlio di Rosetta . .	2469 ,,
Passo Venezia . . .	2298 ,,

According to Baedeker's " Guide to The Eastern Alps " (1888), most of these mountains have been ascended since the first edition of this book was published. The Pala di San Martino was first ascended in 1878 by Herr Meurer and the Marchese Pallavicini.—A. B. E.

* Mr. F. F. Tuckett has pointed out to me that a small, and apparently a permanent, accumulation of ice, scarcely to be dignified by the name of a glacier, is to be found near the Pala di San Martino. (*Note to Second Edition.*)

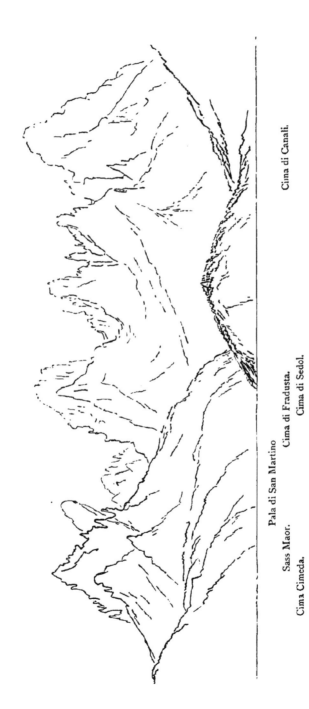

Sass Maor.   Pala di San Martino   Cima di Fradusta.

Cima Cimeda.   Cima di Sedol.   Cima di Canali.

PANORAMIC VIEW OF VAL DI CANALI.

What may be the actual diameter of these strange holes, I am unable to guess ; but they look as clean cut, and about as large, as a shot-hole made by a large cannon-ball. Anyone who has ever visited the valley of Grindelwald will remember a similar orifice, locally known as Martinsloch, in the crest of the Eigher.

Waiting here only long enough to get the accompanying outline of the range as seen from Count Welsperg's meadow, we again push on ; for clouds are already beginning to gather about the summit of the Cima di Canali, and we are still far from the head of the valley. Hence the path lies for a long way in the shade of the fir-woods ; then by the side of the torrent bed—here very wide, and bordered by a broad tract of glaring white stones ; then through more woods, with openings here and there through which the great mountains are seen to be ever closing in, nearer and loftier. For the farther one penetrates up this wonderful glen the more overwhelming is the effect, till the whole culminates at last in a scene of savage grandeur unsurpassed, if I may venture to say so, by even the great *impasse* at Macugnana.

By the time we reach this ultimate point, however, the rapid mists have already gathered in a way which, though it enhances the mystery and sublimity of the view, is yet sufficiently disappointing at the end of more than three hours' journey. The Sasso di Campo, which we are destined never to see clearly, is so shrouded in dense vapours that only the lower flanks of it are seen reaching up into the gloom. The huge Cima di Canali, visible less than an hour ago, towers overhead, already half lost in a heavy grey cloud. A long serrated line of stony Col uniting these two great masses, shows all striated and ribbed by the action of pre-historic glaciers. Green pastures, and above these, dark fir-woods, climb to about one third of the height of the Cima di Canali; while innumerable threads of white waterfall are seen leaping from ledge to ledge and wavering down the cliffs in every direction. These waters, gathered into three roaring torrents, hence rush down from three different points, and unite somewhat lower in one broad impetuous stream. The sound of them fills the air like the roaring of the sea upon an ironbound coast. The fir-trees shiver, as if a storm were at hand. I doubt if a more lonely, desolate, and tremendous scene is to be found this side of the Andes.

So many interesting excursions may be made from Primiero, that the traveller who has only two or three days to dispose of cannot hope to achieve even the half of them, The place, indeed, is one to be chosen for a lengthened sojourn, and treated as headquarters till the neighbourhood is exhausted. We regretted at the time that it was not in our power to do so. The ascent of Monte Pavione (an uncommon looking mountain, in shape like a stunted pyramid, lying away to the S. of Primiero, and forming the highest point of the range known as the Vette di Feltre) is said not to be difficult. The view from the summit commands the whole sweep of the Adriatic coast from the mouth of the Isonzo at the head of the Gulf of Trieste on the one side, to Chioggia, twenty miles south of Venice on the other. Many rarest plants are also to be found on the mountain,

amongst which the following are enumerated by Ball :—
*Anemone baldensis, Anemone narcissiflora, Ranunculus
Seguieri* and *Ranunculus Thora, Delphinium montanum,
Papaver pyrenaicum, Arabis pumila, Alyssum Wulfenia-
num, Cochlearia brevicaulis, Alsine lanceolata, Alsine
graminifolia, Cerastium tomentosum, Phaca frigida, Poten-
tilla nitida, Saxifraga petræa, Valeriana elongata, Ptar-
mica oxyloba, Scorzonera purpurea, Pæderota Ageria,
Pæderota Bonarota, Pedicularis rosea, Primula Facchinii,
Cortusa Matthioli, Avena Hostii,* and *Asplenium Seelosii.*
This excursion involves a night, and a hay-bed in a
châlet on the Agnerola Alp at the foot of the Pavione
rocks; but this is a difficulty that would not have
deterred us, had we been travelling in a larger party.

The ascent of Monte Arzon, a mountain rising about
8,700 feet, and situated in a fine central position about
three miles N.W. of Primiero, is also strongly recommended by the local guides.

A very interesting excursion, however, and one which
can be accomplished all the way on mules, is to the
Ponte dello Schios on Monte Vederne, a small wooded
mountain bordering the west bank of the Cismone,
about three miles below Primiero. The way thither
lies along the main road as far as the villages of
Mezzano and Imer; thence over the Cismone bridge,
and up a rough caretta track all black underfoot from
charcoal droppings, which skirts the pine-slopes overhanging the gorge of the Noana. The path rises and
winds continuously. The Primiero valley is left behind
and soon lost to sight. The torrent down below
becomes inaudible. We meet a train of mules laden

with huge black sacks of charcoal, and have to back up against the rock to let them pass. They, however, according to the nature of mules, prefer the brink of the precipice, and pick their way past with half their bulky burdens overhanging the abyss.

At length, when we have mounted to the height of perhaps fifteen hundred feet above the valley, we pass under an impending roof of rock, and find ourselves at the mouth of a gigantic cavern which looks as if it might have been scooped out by some mighty water-power ages ago, when the world was yet unfinished. Beyond this cavern there rises a semi-circular wall of vertical precipice, at the end of which a small cascade leaps out over the ledge and is dispersed in mist before it reaches the brown pool below. Our path turns abruptly into and round the inside of the cavern, and then along a giddy wooden shelf supported on pine-trunks driven into the face of the rock-wall opposite. This is the Ponte dello Schios. The shelf looks horribly unsafe, but is extremely picturesque; and the whole scene, though on a grander scale, reminds one of the cavern and wooden gallery at Tivoli. A little carved and painted Christ under a pent-house roof is fixed against the rock, just at the beginning of the bridge; and an old white-haired man coming down that way, pulls off his hat and stays to mutter an Ave as we pass.

From this point, a short ascent of about another thousand feet would bring us out, we are told, upon the Agnerola Alp; but we dare go no farther, for the sun is already near setting, and we fear to be overtaken by the dusk. Still it is none the less tantalising to find that we have made nearly one third of the ascent to Monte Pavione without knowing it.

Leaving Primiero for Predazzo . . . . but stay ; how can I leave Primiero without one word of Signor Prospero ?—Signor Prospero, genial, fussy, courteous, enthusiastic, indefatigable, voluble ; Signor Prospero, whose glory it is to be a member of the Italian Club Alpino ; who believes the British nation to be the most enlightened that the sun shines upon ; who so worships the very name of Ball and Leslie Stephen that he all but takes his hat off when he mentions them, as if they were his patron saints ; who vaguely imagines that every English tourist must be in some way or other illustrious ; that all our autographs are worth having ; and that the universal family of Smith represents the flower of the human race !

Shall I ever forget that blazing afternoon when, gaitered, white-hatted, his garments buttoned all awry, and a striped silk umbrella under his arm, he escorted me to Signor Sartoris's museum and apiary ?—or that evening when he came to call, and we entertained him on the landing, and he talked for two hours without stopping about State Education, the Darwin theory, the Calculating Machine, Capital punishment, Prehistoric Man, the Atlantic cable, Universal Suffrage, Positivism, the Solar Spectrum, the Alabama claims, the sources of the Nile, the Prussian military system, Liberty of the Press, the Armstrong gun, the Suez canal, the Eruption of Vesuvius, and the Rights of Women ! A kindly, benevolent, public-spirited old man, eager to promote something like culture and progress in his native town, and interested in all that stirs the great outer world beyond his ken! To establish a more rapid system of postal communication, to get the wire brought over from Feltre, to improve the teaching in the Primiero schools, and to found a local newspaper—these are among the dreams that he is striving to realise. The little Teatro Sociale (for Primiero has its tiny amateur theatre and corps dramatique) is of his creation, and under his management. The new road to Predazzo would not have been put in hand, probably, for the next ten years, but for the energy with which he was continually agitating the question in Primiero.

" Ecco, Signora," he said, unconsciously quoting the dying words of Goethe, " what we want in our little valley is *more light*. Our people are not poor, but they dwell in the darkness of ignorance. We have schools for the children, it is true ; but then what is to be done with their parents who regard geography as an invention of—*con rispetta*—the Devil ! "

I think it was that same evening, when all the lamps were out and the little world of Primiero had well-nigh dropped into its first sound sleep, that we heard a delicious tenor, rich and sweet and powerful, ring out suddenly through the silence of the night. It began at a little distance off—died away—came back again—then ceased close under our windows. The air was Verdi's, hackneyed and commonplace enough ; but the voice was fresh and faultless, and belonged, as we learnt next day, to young Bonetti, the second son of our landlady. He told us that his name was already entered on the

R

books of the Conservatoire of Milan, and that he was to begin his vocal studies in November. It is said that so fine a voice has not been heard within the walls of the Academy for more than a quarter of a century.

With regard to Signor Sartoris, just named, he seems to have raised Apiculture to the dignity of a science. Self-taught, he has discovered how to regulate the productiveness of the race, and is said to be able, unhurt and unstung, to take in his hand and transfer from hive to hive the Queen-bee and her court. How far this may be true I cannot say; but I saw his museum and his apiary—the former a collection of all the bees, beetles, butterflies, woods, minerals, and chemical products of the district—the other a Ghetto of hives, one hundred and fifty in number, containing a population of several millions of bees, the whole packed into a tiny back-garden less than an eighth of an acre in extent. His father and sister show these things with pardonable pride; but Signor Sartoris no longer lives in Primiero. Though not yet thirty years of age, he has been appointed Director of a Government apiary at Milan, and is there developing his system with extraordinary success.

And now we must say farewell to Primiero and all its notabilities; we must say farewell and be going again, for there are yet many places to be seen and many miles to be traversed, and the pleasantest tours and the brightest summers cannot last for ever. So away we ride again, one bright early morning, overwhelmed with good wishes and kind offices, and presented by Signora Bonetti with a parting testimonial in the form of a big cake—so big that it can hardly be got into the basket.

Our way lies by the new military road as far as it is yet completed,* and along the Val Cismone—that great valley which descends from the north-west, running parallel with the Val Pravitale, and divided from it by the range that ends with the Cima Cimeda. Following almost the same course at first as the old road, and crossing the stream near Siror, (where may yet be seen the entrance to the ancient silver-mine) the new "strada" then strikes up in a series of bold zigzags, and is carried at a great height along the precipitous slopes bordering the west bank of the torrent. Up here, all is silent, all is solitary. A couple of Austrian gendarmes—a little group of cantonniers at work upon the road—a tiny donkey staggering under a gigantic load of hay; these are all the living things we meet for hours. But the great mountains on the opposite side of the valley keep us solemn company during many a mile—a wonderful chain of Dolomite peaks, less incredible in outline, perhaps, than those of the Val di Canali,† but rising to a more uniformly lofty elevation.

One by one, we pass them in review. First comes the Cima Cimeda, called by Mr. Gilbert the Procession

* Now completed. *See foot-note, p. 257.*

† The relative position of these mountain ranges can only be understood by reference to the map, where it will be seen that the line of peaks which ends with the Cima Cimedo on the S., and leads up to the Cimon della Pala on the N., does not, in point of fact, form any part of the boundaries of the Val di Canali, but, on the contrary, walls in the W. side of the Val Pravitale. This range is very imperfectly seen from Primiero, and still more imperfectly from the Val di Canali, the perspective in both instances being so abrupt as only to show the peaks in line, one behind the other.

R 2

Mountain, but, to my thinking, more like some strange petrified sea-monster bristling all over with gigantic feelers ; next come the mighty leaning towers of the Sass Maor ; then the Cima Cimerla (so called from the Cimerla woods below), the Cima Pravitale, and the Cima di Ball, three names as yet not entered in the maps ; lastly the vast perpendicular wall of the Pala di San Martino, which rises grander and steeper with every foot of the road, and seems to fill the scene. At length, however, we turn away from this great panorama, through a pine-wood and across a green undulating Alp all ablaze with gorgeous golden lilies ; and so arrive at the tiny church and rambling Hospice of San Martino.*

Arriving here after four hours of easy riding, we pause to take half-an-hour's rest before attacking the Costonzella pass. It is a large, dirty, ruinous place— once a monastery ; then a feudal residence ; now an inn and farm-house combined. It was built somewhere about the middle of the eleventh century, while Edward the Confessor was yet reigning here in England, and when the Bishops of Trent were lords of Primiero. It was these spiritual rulers who erected the church, the monastery, and the Hospice, and dedicated them to San Martino.

Having ordered coffee, we are shown up into a big upper room at the end of a wilderness of passages. It has been a grand room once upon a time—perhaps the

---

* A new hotel has now been added to the Hospice building, and is much frequented during the summer. San Martino is finely situate and stands at a much greater altitude than Cortina. (*Note to Second Edition.*)

prior's own snuggery; perhaps a guest-chamber for travellers of distinction. The walls and ceiling are all of oak, panelled in sunk squares ornamented with bosses and richly carved. A carved shield charged with the Welsperg arms in faded gold and colours commemorates the time when the building had ceased to be a monastery and became a baronial residence. Old family portraits of dead-and-gone Welspergs hang all awry upon the walls and stand piled in corners, draped in cobwebs and loaded with the dust of years—courtiers in flowing wigs, prelates in lace, doughty commanders in shining cuirasses. A certain " Princess Canonicus " in a religious dress, with long white hands that Vandyke might almost have painted, must have been pretty in her day, if the old limner did not flatter her. These by-gone lords and ladies, together with a curious old porcelain stove in blue and white Delft, two squalid beds, a deal table, and four straw-bottomed chairs, are all the furniture the room contains. It ought to be a haunted chamber, and is the very place in which to lay the scene of a ghost story. The whole house, indeed, has a fine murderous look about it, and is as solitary, forlorn, and mediæval a place as any sensation novelist could desire for a mise en scène.

The good road ends at San Martino ; that is to say, it extends in an unfinished, impassable state for another two or three miles ; but we strike straight up the Col by a bridle-path * leading up a wild glen and over a grassy

* This old bridle-path is still to be preferred by pedestrians to the long zig-zags and windings of the new road, now completed. *See foot-note, p.* 257. (*Note to Second Edition.*)

slope thick with crimson Alp-roses, till all at once we find ourselves on the summit of the pass, standing just below the base of the Cimon della Pala. The air up here is cold and rare. The pass rises to a height of 6,657 feet; the stupendous Dolomite wall above our heads towers up to 11,000 feet, of which more than 3,000 feet are sheer, overhanging precipice. In form it is like a gigantic headstone, with a pyramidal coping-stone on the top. Terrific vertical fissures which look as if ready to gape and fall apart at any moment, give a frightful appearance of insecurity to the whole mass. Not the Matterhorn itself, for all its cruel look and tragic story, impresses one with such a sense of danger, and such a feeling of one's own smallness and helplessness, as the Cimon della Pala.

Looking back from this elevation in the direction of Primiero, we get a wonderful view of the Pala di San Martino, the Sass Maor, and the summits of the Val di Canali; beyond these, the Pavione and the Vette di Feltre; and beyond these again, a vast troubled sea of pale blue and violet peaks, some of which encompass the lake of Garda, while some watch over the towers of Verona.

And now the clouds, which for the last hour or two have been gathering at our heels, begin driving up the pass and scudding across the face of the great Dolomites. Soon all the lower summits are obscured; the vapours roll up in angry masses; and the huge peaks now vanish, now look out fitfully, in gloom and storm-cloud.

Passing an unfinished building (presumably a new

Hospice) on the top of the pass, we emerge upon the Costonzella Alp. Here an entirely new panorama is unfolded before our eyes. The great prairie undulates away to a vast distance underfoot; to the North opens another sea of peaks terminating with the summits beyond Innsbruck; to the East lie wooded hills and rich pasturages; to the West a steep descent of apparently interminable pine-forest bounded by a new range of dark, low, purple peaks streaked here and there with snow. The loftiest and nearest of these is the Monte Colbricon. It needs no geological knowledge to see at once that these new mountains are not Dolomite; or that we are, in fact, entering upon the first outlying porphyries of Predazzo.

The path now turns abruptly to the left, and plunges down through the steep pine forest. Somewhere among those green abysses, half-way between here and Predazzo, lies the Hospice of Paneveggio, where we are to dine and take our mid-day rest. On the verge of the dip we dismount, promising ourselves to walk so far, and leaving the men and mules to follow. It is a grand forest. The primeval pines up here are of gigantic size, rising from eighty to over a hundred feet, enormous in girth, and garlanded with hoary grey-green moss, the growth of centuries. Except only the pines close under the summit of the Wengern Alp on the Grindelwald side, I have never seen any so ancient and so majestic. As we descend they become smaller, and after the first five or six hundred feet, they dwindle to the average size.

A fairly good path, cool and shady, carried down for a distance of more than 1500 feet in a series of bold zigzags,* and commanding here and there grand sweeping views of forest slope and valley, brings us at the end of two hours' rapid walking to an open space of green pasture, in the midst of which are clustered a wee church, a pretty white hostelry, and a group of picturesque farm-buildings. Steep hillsides of pine-woods enclose this little nest on every side. There is a pleasant sound of running water, and a tinkling of cow-bells, on the air. The hay-makers on the grassy slope behind the house are singing at their work—singing what sounds like an old German chorale, in four parts. It is a delicious place; so peaceful, so pastoral, so clean, that we are almost tempted to change our plans, and stay here till to-morrow.

By and by, however, when the two hours have expired and the mules are brought round, we go on again, though regretfully. At this point, we enter the Val Travignolo; here only a deep torrent-gorge between steep woods, but broadening out by and by into corn-fields and pasture meadows rich in all kinds of wild lilies, orange, and silver-white, and pinky turkscaps speckled with dull crimson. Thus, always descending, and overtaken every now and then by light showers followed by bursts of fleeting sunshine, we arrive, at the end of nearly three more hours, in sight of Predazzo, a widely scattered village in a green basin at the end of the valley. It looks like a prosperous place. The houses are large and substantial, with jutting Tyrolean eaves. Two church spires rise high above the clustered roofs. Farm-buildings and Swiss-looking brown châlets

---

* Pedestrians may avoid the zig-zags by following the telegraph poles.

are scattered over the green slopes that circle round the town ; and as we draw nearer, we find ourselves traversing an extensive suburb of saw-mills and timber yards, which here skirt both banks of the torrent.

And now—following at the tail of a long procession of grave, cream-coloured cows, all shod like horses with iron shoes, and carrying enormous bells about their necks —we make our entry into the town. The children run out into the road and shout at our approach. The elder folks come to their house-doors and stare in silence. The Austrian gendarme at the door of the guard-house lifts two fingers to the side of his cap in military fashion as we pass. Then, emerging upon an open space of scattered houses surrounding the two churches, we find ourselves at the door of a large, old-fashioned, many-windowed inn, the very counterpart of the ancient "Stern" at Innsbruck, over the arched entrance to which swings a gilded ship—the sign of the Nave d'Oro.

THE FASSA THAL AND THE FEDAJA PASS.

A VILLAGE IN A CRATER—PREDAZZO AND ITS COMMERCE—PROSPERITY *VERSUS* PICTURESQUENESS—FOOTSTEPS OF THE ETRUSCANS—THE VAL D'AVISIO—MOENA—THE PORPHYRY OF THE FASSA THAL—VIGO AND THE FAT MAIDEN — CAMPIDELLO — MONTE VERNALE — THE GORGE OF THE AVISIO—THE FEDAJA ALP AND THE FEDAJA LAKE—THE GORGE OF SOTTOGUDA AGAIN—HOME TO CAPRILE.

CHAPTER XI.

THE FASSA THAL AND THE FEDAJA PASS.

THE most unscientific observer sees at a first glance
that the lakes of Albano and Nemi occupy the craters
of extinct volcanoes. The craters are there, cup-like,
distinct, and tell their own story. You must climb a
mountain-side to get to the level of them. You stand
on the rim of one ; you look down into it ; you walk all
round it ; or you may descend to the water-level at the
bottom. Nothing can be clearer, or more satisfactory.
But it is startling to be told that Predazzo occupies just
such an extinct crater, and that the mountains which
hem it in on all sides—the Monte Mulat, the Monte
Viesena, the Weisshorn and others—consist of igneous
rock thrown up, lava-like, from that ancient centre at
some incalculably remote period of geologic history.
For here is neither cone, nor mountain, nor amphitheatre of convergent slopes ; nothing, in short, in the
appearance of either the alluvial flat or the surrounding
heights which may at all correspond to one's precon-
ceived ideas of volcanic scenery.

Yet here, as we are told by Richthofen and others,

there must once have been a great eruptive centre, breaking out again and again, and each time throwing up a different kind of rock :—first Syenite ; then Tourmaline granite ; then Uralite porphyry ; then melaphyr ; then, last of all, porphyrite, and the unique Syenite porphyry, famous for its crystals, and unknown elsewhere.*

It is this great variety in the material of the Predazzo rocks, and the immense mineralogical wealth consequent upon this variety, that have for more than a century attracted hither so many men of science from all parts of Europe.

The town—now quiet enough, except as regards its commercial activity—is said to occupy the centre of the ancient crater. It stands, at all events, midway between Cavalese and Moena, just at the junction of the Fiemme or Fleims valley with the Val Travignolo. It is a very prosperous place. The people, though an Italian-speaking race, are wholly Austrian in their sympathies, and are supposed to come chiefly of a Teutonic stock. They are particularly intelligent, industrious, and energetic. They have a fertile valley which they know how to cultivate, and mountains rich in mineral products which they are rapidly and successfully developing. As iron-masters, as hay-merchants, as wood-contractors, they carry on an extensive Northern trade, and travel annually for purposes of commerce in

---

* For a brief and intelligible account of the geology of the Fiemme, Fassa, Ampezzo, and other S. Tyrolean valleys, I cannot do better than refer the reader to Mr. G. C. Churchill's " Physical Description of the Dolomite Regions," which forms the concluding chapter of "The Dolomite Mountains."

Germany, Hungary, Transylvania, and Switzerland. Large iron-foundries and long lines of busy saw-mills give an unwonted air of activity to the place. New works, new yards, new and substantial dwelling-houses, are rapidly springing up in every direction. A new Gothic church with a smart roof of gaily coloured tiles, red, green, and yellow, has lately been erected on the South side of the village and there become the centre of an increasing suburb. The schools are said to be excellent; and a well-informed priest from whom I learned most of the foregoing particulars, said the children were full of spirit and intelligence. He also told me that there were now no noble families in Predazzo; but only a wealthy territorial and commercial middle class. He estimated the gross population of the Commune at something over 3000 souls.*

Prosperity and picturesqueness, however, are not wont to travel hand in hand; and it must be admitted that these foundries and timber-yards by no means add to the pastoral beauty of the valley. They spoil it for the artist, just as the mills and factories of the last twenty years have spoiled the once romantic valley of Glarus in Switzerland. Still, down among the wooden houses in the old part of the village, where the women wash their vegetables and fill their pitchers at the stone fountain in the middle of the street, some quaint Prout-like subjects may yet be found. The old

* In 1888 Baedeker gives the population of Predazzo as only 3335, thus showing that the place has grown but little since the above was written. (*Note to Second Edition.*)

church, with its characteristic Tyrolese belfry and steep gable-roof, is charmingly mediæval; and the view from the meadows at the back of the Nave d'Oro, bringing in the two churches and looking straight up the Val Travignolo to where the Cimon della Pala and the Cima della Vezzana tower up

PREDAZZO.

against the distant horizon, seemed to me quite worth a careful sketch.

While I was making the sketch—sitting in the shade of a little shrine among the field-paths—two Austrian soldiers came by, and stayed to look on. They were simple, friendly fellows, natives of Trient, and quartered, they said, with three others of their regiment, in Predazzo. Not knowing that they acted in the double capacity of local police and military patrol, I asked what they could find to do in so peaceful a place.

"Nay, Signora," said the one who talked most, "we have the work of ten men upon our hands. Night and day alike, we patrol the woods, roads, and passes for twelve miles in every direction. Our rounds are long and fatiguing—our intervals of rest, very brief. We get but one day's rest in every seven, and one night in every four or five."

I afterwards learned that there were five other soldiers quartered at Cavalese, as many more at Moena, and so on throughout every petty commune; and that, according to the general impression, the men were greatly overworked.

The Nave d'Oro (without disparagement of the inns at either Caprile or Primiero) was undoubtedly the best albergo we came upon during the whole tour. The house is large, clean, and well-furnished; the food excellent; and the accommodation in every way of a superior character. The landlord—Francesco Giacomelli by name—is a sedate, well-informed man, a fair mineralogist and geologist, and proud to tell of the illustrious savants who have from time to time put up at his house and explored the neighbourhood under his guidance. He keeps collections of local minerals for sale, among which the orthoclase crystals struck us as being extraordinarily large and beautiful.

Lying among these crystals, in one of Signor Giacomelli's specimen-cases, the writer observed a small penannular bronze bracelet of Etruscan pattern and

s

very delicate workmanship, coated with the fine green rust of antiquity; and learned on enquiry that it had been discovered with other similar objects in the cutting of a new road near the neighbouring village of Ziano.

The "find" consisted of a sword, a torque, some fibulæ, a number of bronze pins, and several bracelets; all of which, with this one exception, were immediately purchased by a Viennese gentleman who chanced to be staying in Predazzo at the time. It is singular that no vases seem to have been found, and no masonry to indicate that the road-makers had broken into a tomb. It seemed rather as if some warrior had been hastily laid in earth, just as he fell. On the other hand, however, this little bracelet (which, being accidentally mislaid, had escaped the Viennese collector, and so came to be bought for a few francs by myself) was evidently a woman's ornament.

It is interesting to know that like traces of the Northward migration of the Etruscan races when driven by the Gauls from their settlements on the Po, have been found at Matrey, Sonnenburg, and other places of S. Tyrol:—one notable instance being the discovery of an inscribed bronze bucket near the mouth of the Val Di Cembra (which is, in fact, a Westward prolongation of the Fiemme Valley) in 1828. I myself saw, in the little museum of Signor Sartoris at Primiero, a small aryballos-shaped vase of yellow clay with red ornamentation, which I should undoubtedly take to be of Etruscan workmanship, and which they told me had been found by himself in a field not far from the town.

Of the remarkable sepulchral discoveries made at St. Ulrich in the Grödner-Thal, A.D. 1848, and of Herr Pürger's interesting Etruscan objects found in those graves, I shall have to tell farther on.

The Nave d'Oro at Predazzo is a curious old house, and it has belonged to the Giacomelli family for many centuries. The Giacomellis, as I have said elsewhere, were once noble, and their armorial bearings still decorate many of the old carved doorways, ceilings and chimney-pieces of their ancestral home ; but that was long ago, and they have been innkeepers now for more than a century. Their visitor's book is quite a venerable volume, and contains, among the usual irrelevant rubbish of such collections, the handwriting of Humboldt, Fuchs, Richthofen, Sir Roderick Murchison, the Elie de Beaumonts, and other European celebrities. But some nefarious autograph-hunter has abstracted one of the greatest treasures the book contained—the signature of the discoverer of the Georgium Sidus.

Here too, among the latest entries, a certain Dr. Reinhart of Munich had exercised his Latinity in the following pithy sentence :—

"*Viator ! Cave Tabernum Bernhart in Campidello !* "

This ominous caution—so much the more impressive for being so vague—had the effect of deciding us against putting up for a night, or even a midday rest, at the albergo in question. How many travellers since then, I wonder, have like us accepted the good Doctor's salutary warning ? And what would have happened to us if we had neglected it ?

s 2

The Val Fiemme, or Fleims Thal (about the middle of which Predazzo is situate), is but one portion of an immensely long tortuous valley called in part the Val Fassa, in part the Val Fiemme, in part the Val Cembra, which begins with the source of the Avisio in that depression between the Marmolata and the Monte Padon which is known as the Fedaja pass, and ends where the torrent debouches into the Eisack at Lavis, seven miles north of Trient. The collective name for this chain of valleys is the Val d'Avisio ; and, except at quite the upper end of the Fassa division, it is the least picturesque of any that came within the compass of our journey.

Leaving Predazzo after one day of rest—for, however attractive to geologists and mineralogists, it has no excursions to repay the unscientific visitor—we next pursued our course up the valley purposing to put up for a couple of nights at Vigo in the Fassa Thal, and thence to explore the cirque of the Rosengarten, and ascend the Sasso dei Mugoni.*

It is a dull day when we start, having a somewhat dull journey before us. Our way lies at first between a double range of low hills partly clothed with pine-forest, and partly with scrub. These hills, which are of the dark igneous rock thrown up from the Predazzo crater, hide the loftier peaks and are not picturesque at all. By and by comes a long straight road, terminated miles away by the village of Moena. Going along this road, a few unmistakably Dolomitic summits begin to

---

* There is now daily communication by omnibus between Predazzo and Vigo by the road from Cavalese to Vigo. (*Note to Second Edition.*)

peer up here and there above the barren hills to the left; and straight ahead, far beyond Moena, rises the Monte Boé, looking like an immense fort on a grand pedestal of rock, its battlements lost in the clouds. This Monte Boé, the southernmost bastion of the huge Sella Massif, is also known as the Monte Pordoi. It has been ascended by Dr. Grohmann, who calculates its height at 10,341 feet.

Passing through Moena—a large, straggling, wood-cutting village—and crossing a couple of bridges, we leave the high road and strike up a steep mule-path on the opposite bank of the torrent. It is the same valley, and the same water; but here above Moena, it is called the Fassa Thal. Looking back from this higher ground, we get a fine view over the Monte Latemar (8,983 feet) and its far-reaching fir-forests; while the wild peaks of the Rosengarten and Lang Kofel came into sight above the lower slopes of Costa-lunga.

And now, in rich contrast to the pallid Dolomites soaring high in the distance, the famous porphyry of the Fassa Thal begins to break out in crimson patches among the lower hills, and to appear in the cliff-walls that border the Avisio far below. Yonder, where the stream takes a sudden bend, two isolated porphyry pillars jut out on either side, forming a natural portal through which the narrowed waters rush impetuously. A little farther still, and a whole mountain side of the precious marble, quarried terrace above terrace, and apparently of inexhaustible richness, is laid bare to view. Now we recross the stream, and pass through the village of Soraga. Here, everything except the grass and the trees, is crimson. The ploughed fields are crimson; the mud underfoot is crimson; the little torrent hurrying down the ravine by the roadside is crimson; the very puddles are crimson also. Even the roads are mended with porphyry, and great blocks of it lie piled by the wayside, waiting for the hammer of the stonebreaker.

The sky, which has all day been murky, now seems to be coming down lower and lower, like a heavy grey curtain. The air grows chill. A cold leaden tint spreads over the landscape; and the long dull road seems to grow longer and duller the farther we follow it. At length we come in sight of Vigo, a village clustered high upon a hill-side to the left, backed by lofty slopes of fir-forest, down which the gathering mists are creeping fast. A steep path leads up to the village, whence, looking over to the north-east where the horizon is still clear, we catch a momentary end-wise glimpse of the Marmolata.

And now we are overtaken by a smiling lad with a bunch of wild strawberries in his hat, who turns out to be young Rizzi, son of old Rizzi who keeps the albergo up here at Vigo—a large, dark, dreary house, the entrance to which lies through a filthy cart-shed and up a staircase that looks as if it had not been scrubbed for the last half century. Here we are received by the land-lord's daughter, a fat, bouncing, rosy-cheeked damsel of inexhaustible activity and good humour, who does her best to make us welcome. The inn, however, proves to be quite full, with the exception of one big, treble-bedded room with windows looking to east and north, and a ceiling about seven feet from the floor. And we are fortunate to secure even this; for before we have been half an hour in possession of it, there arrives a party of Germans—hungry, noisy mountaineers, regularly got up for work, with ropes, ice-hatchets, and hobnailed boots—for whom beds have to be made on the landing.

A chill, drizzly evening, a supper irregularly served, and boisterous neighbours in the adjoining rooms, caused us, perhaps unjustly, to take a dislike to Vigo. The house, too, was full of foul smells; and a manure-heap in the cow-yard under one of our windows did not help to improve the atmosphere. So when morning came, bringing a sea of white mist that extinguished all the mountain-tops, we decided to start for home as quickly as possible. In vain the fat maiden represented that to-day it would surely rain, and that if we only delayed till to-morrow we should be certain of magnificent views and splendid weather. In vain she exhausted her eloquence to prove the absurdity of our attacking the Fedaja pass in mist and rain. We did not believe that it was going to be wet; we knew we could take the Fedaja again from Caprile any day we chose; and we were determined to go home.

So by half-past six A.M., behold us on the road again, delighted to get away from Vigo, and hoping for a tolerable day.

It is a sweet, fresh morning. The vapours are rolling and rising, the clouds parting, and stray gleams of sunshine gliding now and then across the hill-sides. But the mountain-tops continue to be veiled in masses of soft, white haze, and only thrust a tusk out here and there. Confident, however, of fine weather, we laugh the fat maiden to scorn, and ride on our way exulting.

The valley now grows in beauty at every turn. At Mazin, we come upon a picturesque hamlet with a background of ravine and waterfall, and approaching Campidello, we look out anxiously for the strange Dolomite peaks that overhang the village. The mist is thick; but there they are, gleaming grey and ghostlike. Here, too, is the little albergo against which we have been warned by Dr. Reinhart of Munich. It looks rather pretty; but the sight of two extremely dirty and ill-favoured dwarfs—a man and a woman—who come out upon the balcony to stare at the travellers, quite confirms us in the satisfaction with which we ride past the house.

A little higher up the valley, we reach the villages of Gries and Canazei; and, stopping for only a few minutes at Canazei to feed and water the mules, push on rapidly for the Fedaja. Still the scenery continues to increase in beauty. On the hillsides are corn-slopes, woods, and pastures; in the valley, a rushing stream babbles among tamarisk trees and pines. Soon, a fine pyramidal mountain, black and precipitous on the one side, sheeted with snow on the other, comes into sight at the head of an opening valley to the right. We take it at first for the Marmolata; but it proves to be the Monte Vernale, a less lofty but far more difficult mountain, calculated at 9,845 feet in height.

Now the path turns off to the left, threading the two miserable hamlets of Alba and Penia, and rising rapidly through a grand rocky gorge which gets finer and more savage the higher it climbs. Steep precipices shut it in on the one hand, and barren slopes battlemented with jagged rocks upon the other. The Avisio, here a mere thread of torrent, foams from rock to rock in innumerable tiny cascades. Wide-spreading firs and larches make a green roof overhead, and the path is carpeted with fragrant spines upon which the mules tread noiselessly. Presently we come in sight of a fine waterfall which issues from a fissure in the face of the great cliff to the right, descends in two bold leaps, and vanishes amid the depths of the fir forest below.

The gorge now closes in nearer and steeper, our upward path being indicated by the giddy windings of a little hand-rail which scales the face of a huge rock straight ahead. It is here too steep and slippery for riding, so we dismount and walk.

Alas! the fat maiden was right, after all. The mist which has been lightly drifting in our faces for the last half hour, now sets in with a will, and becomes a steady pour. Drenched and silent, we toil up the stony path and wish ourselves back at Vigo. An hour hence, says Clementi, we shall come to some châlets and cattle-sheds; but there is no Hospice to look forward to here, as on most other passes. By and by, however, where the climb attains its worst pitch of steepness and slipperiness, we pass a succession of little carved and coloured " Stazione " nailed at short intervals against the rock, for the benefit of such pious souls as may care to say a few Aves by the way; and these lead to a tiny chapel not much bigger than a sentry-box, into which we are thankful to creep for temporary shelter. A wretched crucifixion by some village artist, a few faded wild flowers in a broken mug, and a multitude of votive hearts, arms, legs, eyes and so forth, in tinsel and coloured wax, decorate the little altar; while securely embedded in a niche in the wall, chained, padlocked, and iron-bound, there stands a small coffer with a slit in the lid, for the reception of stray soldi.

Here, glad of even a few minutes' respite from the pitiless deluge without, we wring the rain from our dripping garments, and divide with the men what we have left of bread and wine; not forgetting the wet and melancholy mules, who receive a lump of bread apiece, and are comforted by L. with bits of sugar.

It is still pouring when we go on again, and it continues to pour steadily. For full another hour we keep on under these pleasant circumstances, always on foot; and then, quite suddenly, we find ourselves close under the western end of the Marmolata. Invisible till this moment, it now looms out all at once in startling proximity. A great blue wrinkled glacier, reaching down out of the mist like a terrible Hand, grasps the grey rock overhead; while beyond and above it, a vast field of stainless snow slopes up into the clouds, without sign of end or limit.

Turning from this grand spectacle to the rocky shelf we have just reached, we find ourselves in a garden of wild flowers. There were none in the gorge below; none by the path-side coming up; but here they are beautiful and abundant, as if fair Earine had lately passed this way, the flowers following in her track :—

"As she had sow'd them with her odorous foot !"

Wetter than wet through one can hardly be ; so we despatch Clementi up the rock to fetch some bunches of the rare, white, velvety Edelweiss, while we quickly gather such lower plants as grow within easy reach. Thus in the pelting rain we secure some specimens of the *Orobus luteus, Dryas octopetala, Primula farinosa, Pinguicula grandiflora, Cynanchum Vincetoxicum, Orchis nigra*, &c., &c. ; besides several varieties of cyclamen, gentians, and ferns.

Again a little higher, and we reach the summit of the pass—a lonely upper world of rich sward, bounded on the left by the splintered peaks of Monte Padon, and on the right by the lower slopes of the Marmolata, which rises direct from the grassy level on which we stand. This is the Piana Fedaja, or Fedaja Alp. A dozen or so of rough wooden châlets are here clustered together ; mere cattle-refuges and hay-sheds, one of which, being a trifle more air-tight than the rest, is decorated with a coloured Christus over the doorway, and serves as a sleeping place for travellers who are about to make the ascent of the mountain.

The rain now abates somewhat of its violence, and, the way being once more level, riding again becomes practicable. Thus we go on ; a second and a third great glacier creeping into sight as the first is left behind. These each show a brown margin of moraine ; the last glacier being of immense extent, as large apparently as the lower glacier of Grindelwald. While we are yet looking at them, however, a tall, strange, ghost-like mist stalks swiftly across the snow, and veils all but the brown rocks abutting on the pass. In a moment the great mountain has melted away, and we see it no more.*

The Fedaja Alp is just the width of the Marmolata, and no more. It begins with the Western, and ends with the Eastern extremity of the mountain. Here, at the foot of the huge dark rock known as the Piz Seranta, lies an exquisite little dark green tarn surrounded by slopes of crimson Alp-roses. The rain having now ceased for a moment, its waters, ruffled only by the flight of a small brown moor-hen, are as placid as a sheet of green glass.

Another yard or two of rocky path, and we come to an upright, mossy stone bearing an illegible inscription. This is the ancient boundary-stone between Italy and Austria—one of the few divisions left unchanged at the last readjustment of the frontier-line. Half of the Marmolata belongs to the House of Hapsburg, and half to the kingdom of Italy. The line of demarcation

---

* The height of the Marmolata, though proved to exceed that of the Cimon della Pala, is not yet thought to be satisfactorily ascertained. The Austrian "Kataster" measurement gives 11,466 feet; while Dr. Grohmann gives a barometrical elevation of only 11,045 feet. Mr. Ball arrived within a few feet of the summit of the second peak of the Marmolata in 1860; Dr. Grohmann ascended it unsuccessfully in 1862, and made the first ascent of the highest summit in 1864. Mr. Tuckett made the second ascent in 1865, by a new and more direct course, and repeated the achievement in 1869.

NOTE TO SECOND EDITION.—The height of the Marmolata, as given by the latest Austrian survey, is now fixed at 3,359 mètres, or 11,020 English feet.

is ingeniously carried along the topmost ridge of ice and glacier, so that, unless by members of the different European Alpine clubs, it is not very likely to become a disputed territory.

From this point, all is descent. Our way lies along a vast green slope, following the course of the Candiarei torrent, but running for a long distance upon the brink of a ruinous gully partly choked with yet unmelted snow. For the path on the Candiarei side has been lately swept away by a torrent of snow and water from the Marmolata, and the whole mountain slope is here one mass of soft red mud, more slippery than ice, full of pits and fissures, and very difficult. Lower down still, the track lies through rich park-like pastures deep in wild-flowers, so bringing us at last to the upper end of the Sottoguda gorge.

No sooner have we entered the defile than the clouds clear off as if by magic. The sun then bursts out in splendour, lighting up the rocks first on one side and then on the other, according as the ravine winds its narrow way. Our wet garments steam as if hung before a blazing fire. The men take off their coats, and carry them on their alpenstocks to dry. The mules prick up their ears and rub their noses together, as if whispering to each other that there is a scent of home upon the air, and that the old familiar stable cannot surely be far distant.

Nor is it ; for already we have emerged into the Val Pettorina. Those green slopes to the left are the slopes of Monte Migion ; these fir-woods to the right are the woods of Monte Pezza. Presently come the dilapidated hamlets of Sottoguda and Sorara; then Rocca on its hillside; then the familiar path down by the torrent-side and across the wooden bridge; then at last Caprile, where a warm welcome awaits us, a heap of English letters, and rest.

THE SASSO BIANCO.

OROGRAPHY OF THE SASSO BIANCO—ITS PANORAMIC POSITION—ITS
SUPERFICIAL EXTENT—ITS GEOLOGY—ASCENT OF THE MOUNTAIN—AN
EXQUISITE MORNING — ANOTHER SAGRO — THE CORN - ZONE—THE
PEZZÉ PROPERTY—THE WILD-FLOWER ZONE—THE UPPER PASTUR-
AGES—WAITING FOR THE MISTS—THE LAST SLOPE—THE SUMMIT—
THE VIEW TO THE NORTH—THE ZILLERTHAL AND ANTHOLZER ALPS—
THE GROSS VENEDIGER—GLIMPSES ON THE SOUTH SIDE—ESTIMATED
HEIGHT OF THE SASSO BIANCO—THE DESCENT—GRATIFICATION OF
THE NATIVES.

CHAPTER XII.

THE SASSO BIANCO.

"An ill-favoured thing, sir," says Touchstone; "but mine own."

Now I will not say that the Sasso Bianco is an ill-favoured mountain—Heaven forbid! Nor that it is an unimportant mountain; nor even that it is a small mountain. I will not depreciate it at the beginning, in order to rehabilitate it by a coup de théâtre in the end. Neither will I affect to undervalue it for the sake of establishing an ingenious parallel between myself and the Fool.

At the same time, I am anxious not to exaggerate its peculiar qualifications and virtues. For it is with mountain-tops as with other playthings:—having sought to achieve them in the first instance because we value them, we go on valuing them because we have achieved them. We may even admit their ill-favouredness, as Touchstone admits the ill-favouredness of Audrey; but we are apt all the time to over-estimate them in secret —simply because they are our own. I premise therefore that I am not blindly in love with the Sasso Bianco,* and that the following portrait is not flattered.

* Strictly speaking, as I have said elsewhere, the name of Sasso Bianco applies only to the rocky summit of the Monte Pezza.

T

I cannot better describe the Sasso Bianco than by adopting the words of Clementi. It is not a mountain of the first class; but it is high for a mountain of the second class. It is, for instance, 2000 feet, if not 2200 feet, higher than the Rigi, and about 240 feet higher than the Niesen. Its summit stands about 200 feet higher above the lake of Alleghe than the summit of Monte Generoso above the lake of Lugano. It rises considerably above the tree-line, and just falls short of the snow-level. That is to say, we found one unmelted snow-drift about 100 feet below the summit, and there may have been others which we did not see, lurking in inaccessible fissures and crevices. The snow was firm and pure, but the quantity insignificant.

As regards position, I know of no minor Swiss mountain to which I can accurately compare the Sasso Bianco. The Rigi is a mere outlying sentinel, and the view it commands is too distant to be very striking. The same may be said of Monte Generoso, despite its unparalleled panoramic range. The Ægischhorn view is all on one side. The Görner Grat, unrivalled as a near view over snow and ice, is too circumscribed. But the Sasso Bianco stands in the very centre of the Dolomites, like the middle ball upon a Solitaire board, surrounded on all sides by the giants of the district. If one could imagine a fine, detached mountain, clear on all sides, occupying, say, the position of the village of Leuk in the valley of the Rhone, and high enough to command the whole circuit of the Oberland, Monte Rosa, and Mont Blanc ranges, that mountain would fairly represent the kind of position which the Sasso Bianco holds in reference to the scenery by which it is encompassed. I am not acquainted with the view from the Bella Tola in the Valley of the Rhone; but, judging from its situation on the map, it seems just possible that it may supply exactly the parallel of which I am in search.

The mass of Monte Pezza is of considerable extent. Counting from the points locally known as Monte Alto on the West, and Monte Forca on the East, and from the Val Pettorina on the North to the valley of the Biois on the South, it must cover a space of nearly three and a half miles in length by two and a half in breadth. These, of course, are only rough measurements derived partly from personal observation, and partly based upon the Austrian Ordnance Map. In superficial extent as well as in height, the Sasso Bianco (or, more properly, the Monte Pezza) much exceeds the Monte Migion, the Monte Frisolet, and the Monte Fernazza.*

Of the geology of the mountain I am not competent to form an opinion; but according to Ball's geological map, it is composed in part of Porphyry, and in part of Triassic. The light-coloured cliffs of the summit, facing North, (being the part especially designated as the Sasso Bianco) are probably Dolomite. Both in colour and texture the rock appears, at all events, to be of one piece with that of which the great Primiero and Ampezzo peaks are composed.

---

* It is curious that the Monte Fernazza (also known as the Monte Tos) should have been ascended the previous summer by both Mr. F. F. Tuckett and Mr. Gilbert; and that the Sasso Bianco, notwithstanding the much finer view it necessarily commands, should still have escaped.

Of course we decided upon making the ascent almost as soon as we found ourselves back at Caprile. The way up, though long, seemed to be sufficiently easy. There were many paths and char tracks leading from the valley of Alleghe to the farmlands and hamlets scattered along the eastern side of the mountain; but Clementi recommended a path starting from the Val Pettorina, along which we might ride, he said, as far as the highest pastures, and to within about an hour of the summit. As regarded time, he calculated that from four to five hours, including the last hour on foot, would take us from Caprile to the summit.

All this sounded pleasant enough; so it was arranged that Giuseppe should watch the weather, and rouse the household at 3 A.M. whenever a favourable morning should offer. At length, on the morning of the fourth day after our return, the weather being apparently favourable, Giuseppe gave the signal a little before dawn, and by 5 A.M. we were upon our way.

A more lovely morning we have never yet had. The grass, the wild-flowers, the trees, are all drenched with dew and sparkling in the sun. The birds seem wild with delight, and are singing rapturously among the wet green leaves. Crossing the wooden bridge and taking the familiar road up the little Val Pettorina, as if going to Sottoguda, we hear the bells of Rocca ringing high up in the still air, and pass group after group of peasants in their holiday clothes, making for the hill. For it is a festa this bright morning, and the annual Sagro is held at Rocca to-day. Men and women alike pull off their hats as we ride by. All

wish us good morning, and none fail to ask where we
are going.

Turning away presently from the beaten path, we
then strike down to the water's edge, the mules picking
their way along the loose stones bordering the bed of
the Pettorina torrent.   Skirting thus the base of the
hill on which Rocca is built, we cross a higher bridge
and plunge at once into the shade of the firwoods at the
northward base of Monte Pezza.   The path, which is
steep and stony, then winds round to the east, and
brings us out upon a space of cultivated farm-lands just
overhanging the Cordevole.

Here dark firwoods slope in shade down to the valley
below, and higher firwoods climb the mountain-side
above ; while, between both, a belt of green corn-fields,
lighted here and there by fiery sparks of scarlet poppies,
ripples in the breeze and the sunshine.   Peeping up
yonder, just beyond the brink of the woods, rises the
spire of Caprile, while, farther still, a faint ghost of
white vapour soars lazily up from the direction of
Alleghe.   Presently a lark springs out, full-voiced, from
his nest in the barley ; and a troop of children, their
little brown hands full of poppies and corn-flowers,
come chasing each other down the mountain-side.
Such indeed is the idyllic beauty of the whole scene
that even L. (who, with a culpable indifference to glory
which it grieves me to record, was more than half
inclined to stay at home) is moved to admiration, and
admits that, were it to see no more than this, she is
glad to have come.

Meanwhile, we follow a series of narrow footways

winding among fields of young wheat, barley, flax and hemp. Dark Nessol—a confirmed kleptomaniac—grabs huge mouthfuls to right and left, and leaves a trail of devastation behind him. Fair Nessol, on the contrary, looks and longs; but obeying the light hand on his bridle, abstains regretfully.

Presently we leave the fields behind, and mount again into the shade of the forest. Here and there, where the path is very steep, we dismount and walk. Still higher, we emerge upon a zone of rich grass-land full of busy haymakers, and learn that all this part belongs to Signora Pezzé. Twenty-four such pasturages are yet hers; but half the mountain-side belonged to the family in the old times past away.

From this point, and for a long way up, the pasture-land is like a lovely park, rich in grass and interspersed with clumps of firs and larches. As the path rises, however, the trees diminish and the wild-flowers become more abundant. Soon we are in the midst of a terraced garden thick with white and yellow violets, forget-me-nots, great orange and Turkscap lilies, wild sweet-peas, wild sweet-William, and purple Canterbury bells. Here, too, we make acquaintance for the first time with a grotesque, ugly flower bearing a kind of fibrous crest, like a top-knot of spiders' legs. They call it " Capelli di Dio," or God's-Hair. The forget-me-not is here called Fior di Santa Lucia, or Saint Lucy's flower; and the white clover, known only as a wild-flower in South Tyrol, is the Fior di San Giovanni, or Flower of Saint John.

Looking back now towards Monte Migion, I see that

we have long ago overtopped the Sasso di Ronch, which from here looks no bigger than a milestone ; and that we are already higher than the highest ridge of Monte Frisolet. Meanwhile, however, the morning dews keep rising in white vaporous masses from the depths of the valley below, threatening before long to intercept the view. If they should rise to our own level when once we are on the top, as they seem only too likely to do, it is plain that our chances of a panoramic view are lost beyond redemption.

And now the wild-flower zone is left below, and the path, which here circles round a vast amphitheatre in the mountain-side, gets very steep, and strikes up towards the last pasturages. Steep as it is, however, and hewn in places out of the slippery rock, the farmers have for centuries contrived to drag their rough carettini up and down, when the highest hay is gathered. The rock is even worn into deep ruts, just as the pavement of the Via Triumphalis is channelled by Roman chariot-wheels, where it climbs the steep verge of Monte Cavo.

Here the mules scramble on first, and, reaching the green level above, set off on their own account. In vain Clementi runs and shouts after them. They trot resolutely on, till, reaching a little hollow among bushes and deep grass, they bury their noses in a cool rill which they had scented from afar off.

Clementi, coming up red and breathless, wrenches their heads out of the water, and overwhelms them with reproaches. " Holy Mother ! what do they mean by not minding when they are spoken to ? Holy

Mother! what do they mean by drinking cold water when they are as hot as two hot cakes in an oven? Sacramento! Do they want to fall ill and die, out of mere spite towards a master who loves them? Eh, Long-ears! are they deaf? Eh, monsters of mules! do they not understand Italian?"

It is a long, grassy, trough-shaped plateau, with a few gnarled, bloodless old pines scattered about, and two or three tumble-down chalets. Here the char track ends; but we take the mules on a good way farther still, up a steep pitch at the far end of the pasture Alp, and out at last upon a broad ridge terminated towards the North-East by a long slope and an upright wall of rock, like a line of fortification. To right and left, this ridge dips away into unfathomable chasms of misty valley; to the South-West, it runs down to join the great woods which clothe all the Western mass of Monte Pezza. There is nothing, in short, above the point we have now reached, save the slope leading to the summit.

But where is the summit? Seeing us look eagerly towards the rock wall up above, Clementi laughs and shakes his head.

"Ah, no, Signoras," he says. "Non ancora. We must leave the mules here; but from this point we have an hour's walking before us. The Cima is yonder—yonder; seven or eight hundred feet higher!"

It proves, however, to be over a thousand.

The mists, alas! are now swirling up on this side with frightful rapidity. The Val Pettorina and all the Sottoguda side are hidden by the slope above; but the

Val d'Alleghe, the Civetta, and all the peaks lying to the South-West of our position are now visible only in snatches, as the vapours drift and part. The Val Biois, looking over towards Cencenighe and the Cima di Pape, is like a huge caldron sending up volumes of swift steam.

To go on at present is obviously useless; so we make arm-chairs of the saddles and rest awhile upon the grass, while the mules graze, and the men, who have had more than four hours' climbing, light their cigars and lie down in the shade of a big boulder.

Up here, we are already above the tree-level. Glowing Alp-roses and dark blue gentians abound; but the grass all about grows thin and hungerly. According to the aneroid, and without allowing anything for corrections, we have already left Caprile more than 3,500 feet below. That is to say, we have attained an elevation 200 feet higher than the Fedaja pass, and between 20 and 30 feet higher than the Tre Sassi pass, where it will be remembered we reached the snow-level.

Half an hour is consumed thus, in calculating heights, examining maps, and watching the progress of the mists. Sometimes the sun breaks through, and then they part for a moment and drive off in rolling masses. Sometimes they rush up, as if chased before the wind, sweeping all across the ridge, blinding us in white fog, and leaving a clinging damp behind them. At length we decide to push on for the summit. Clementi, who knows the climate, thinks it may clear off at midday, and that we may as well be upon the spot to take advantage of any sudden change for the better. It is now 10.20 A.M., and we have an hour's climbing before us.

Meanwhile, a little lad who has been picked up on the way is left in charge of the mules, with strict injunctions not to let them stray near the edge of the precipice on either side ;—a duty which he fulfils by immediately lying down upon his face in the damp grass, and falling sound asleep.

So we go on again, slowly but steadily, up the long slope and on to the foot of the rock-wall aforesaid. Here are no steps ready hewn. We have to get up as best we can, and the getting up is not easy. The little crevices and inequalities which serve as foot-holes are in places so far apart that it is like going up the steps of the Great Pyramid ; and but for Giuseppe, who goes first in order to do duty as a kind of windlass, the writer, for one, would certainly never have surmounted the barrier.

This stiff little bit over, we expect to see some sign of the summit ; but on the contrary find ourselves, apparently, as far from it as ever. A second and a third slope still rise up ahead, as barren and unpromising as the last.

And now even the Alp-rose has disappeared, and not a bush of any kind breaks the monotony of the surface. But the gentians make a blue carpet underfoot ; and the Edelweiss, so rare elsewhere, so highly prized, flourishes in lavish luxuriance, like a mere weed. Presently we pass an unmelted snowdrift in a hollow some little way below the summit. Then, quite suddenly, a whole army of distant peaks begins to start into sight ; and so, after six hours, we all at once find our-selves upon the top !

We might, of course, have had a better day ; but it is some reward after long toil to find the view to North and West quite free from mist. The vapours are still boiling up in the South and South-East, but not perhaps quite so persistently as an hour ago. At all events they part from time to time, so that in the end, by dint of patient watching, we see all the near peaks in those quarters.

It is now nearly half-past eleven o'clock, and, having eaten nothing since five, we are all as hungry as people have a right to be at an altitude of between four and five thousand feet above the breakfast table. So before attempting to verify peaks, or heights, or relative distances of any kind, we call for the luncheon-basket and turn with undiminished gusto to the familiar meal of hard-boiled eggs and bread. The water in the flask being flat, Clementi fetches up a great lump of snow, and this, melted in the sun and mixed with a little brandy, makes a delicious draught as cold as ice itself.

In the midst of this frugal festivity, Giuseppe, with the keen eye of a chamois-hunter, recognises L.'s maid (whom he calls the " Signora Cameriera ") on the Cordevole bridge just outside the village. We see only a tiny black speck, no bigger than a pin's head ; but Clementi goes so far as to depose to her parasol. In a moment both the men are up, tying a pocket-handkerchief to a white umbrella, and lashing the umbrella upon an

Alpenstock, which they erect for a signal; and the excitement caused by this incident does not subside till the black speck, after remaining stationary upon the bridge for about a quarter of an hour, creeps slowly away and is lost to sight in the direction of Caprile.

Luncheon over, we set to work with maps and field-glasses, to identify all that is visible of the panorama.

We are sitting now on the brink of the great yellowish cliffs which the writer sketched a little while ago from below the Sasso di Ronch.* All the heights and valleys on this side lie spread out before us, like the surface of a relief-map. We look down upon Monte Migion and Monte Frisolet—both green to the top, and scattered over with hamlets, farms, cultivated fields, and fir-forests. Monte Migion, estimated by Trinker at 7,838 feet, lies full 400 feet below; and Monte Frisolet considerably lower still. The Val Pettorina opens just under our feet, and one could almost drop a stone down into the little piazza of Rocca, where the Sagro is going on merrily. We can see the peasants moving to and fro between the church and a great white booth on the top of which a red flag is flying. Now and then, when the wind comes up this way, it brings faint echoes of the bells, and of the braying of a brass band. As for the holiday-folk, they look exactly like a swarm of very small black insects, all in motion. Monte Fernazza, farther to the right, appears to be considerably lower than Monte Migion, but not so low as Monte Frisolet. Except for a blackish ridge of igneous rock cropping out on the side of the pass of Alleghe, this mountain is

* See woodcut, p. 223—"The Sasso Bianco."

green and cultivated like the others, and is apparently about 6,500 feet in height. So much for the minor mountains in our immediate neighbourhood.

Of the larger, the two nearest (each being distant about two miles in a direct line) are the Marmolata and the Civetta. The last fills all the South-Eastern division of the horizon. Large masses of vapour flit from time to time across the face of that vast, fretted screen; but they flit, and pass away, and it lifts its noble head continually into the clear blue depths of the upper sky. The Marmolata stands up in bold profile, undimmed by even a thread of vapour. Mr. Gilbert, seeing this mountain from the Sasso di Dam and getting it also in profile, though from the Western end, compared it to a huge stationery case, its vertical side to the South, and its long snow-slope to the North. But taken here from the East end,* whence one more clearly sees the sharp depression, or couloir, that divides the peaks, it absurdly resembles the familiar cocked hat worn by the first Napoleon, the precipitous side being of course the front of the hat, and the snow-slope corresponding to the back. A great stream of snow lies in the cleft of the couloir, and all the northward slope is outlined, as it seems, in frosted silver; but the great glaciers and snow-fields that lie towards the Fedaja are from here invisible.

The green threshold of the Fedaja pass, and the low

* The author's sketch of the Marmolata from the Pass of Alleghe (p. 333), though taken from a point about two miles to the S.E. of the Sasso Bianco, shows something of this form, although the mountain from that point was much less foreshortened, which diminished the resemblance.

jagged ridge of Monte Padon, rise just North of the extreme Eastern end of the Marmolata, which is buttressed on this side by the black precipices of Seranta. Monte Vernale, repeating from here as from Canazei its curious resemblance to the Marmolata, lurks close under the Southward wall of its huge neighbour, being divided from it by only a little green slope considerably higher than the Fedaja pass, which Clementi points out as the Forcella di Contrin (9,052 feet), and which is also known as the Forcella di Val Ombretta, and as the Passo di Val Fredda. Still lower down towards the South-West lies the Sasso di Val Fredda, still unascended; a little beyond it comes the Monte Ricobetta, locally known as the Monzon, 8,634 feet in height; and on the same parallel, but still farther West, Monte Latemar, on whose summit the vapours rest all day.

North-West of the Marmolata, about nine miles distant as the crow flies, rise the snow-streaked bastions of the Sella Massif, of which, however, only two great towers—the Boé and the Campolungo Spitz—are seen from this side; while in an opening between the Boé and the Marmolata rises a noble, solitary rock which proves to be the Lang Kofel, 10,392 feet in height, and distant about thirteen English miles. A tiny glimpse of the Rosengarten is also seen in the gap above the Forcella di Contrin.

Returning now to the point from which we started, and looking due North straight over the top of Monte Migion, the pinky snow-streaked line of the Sett Sass, divided from Monte Lagazuoi by the Valparola pass, comes into view. The Sasso d'Istria, which looked so imposing from near by, here shows as a small pyramidal rock of no importance ; the castellated crest of Monte Nuvolau dwindles to a tiny ridge on a long green slope ; the caretta track of the Tre Sassi pass winds between both like a white thread ; and Monte Tofana, sulky and cloud-capped, as usual, shows its pyramidal front only once, when the mists roll apart for a few moments.

Following the parallel of the Tofana, we get misty glimpses of the Cristallino peaks, of the Cristallo, of the Drei Zinnen, the Sorapis, and the Croda Malcora. The Rochetta, and the fantastic ridge of the Bec di Mezzodi divide them off like a fence ; while straight away to the East, the Pelmo shows every now and then, quite clear from base to summit. Between the Pelmo and the Croda Malcora, part of the range of the Marmarole, and the curved prow of the Antelao, peep out through window-like openings in the clouds.

Finally, above and beyond all these, ranging from North-West to North-East, in the only direction where the horizon is permanently clear, we look over towards a sea of very distant peaks reaching far away into the heart of Northern Tyrol. To the N.N.W., a little above and to the left of the Sett Sass ridge, we recognise by help of the map the highest summits of the Zillerthal Alps:—the Fuss Stein near the Brenner pass, 11,451 feet in height; the five peaks of the Hornspitzen, ranging from 10,333 feet to 10,842 feet; and the Hochfeiler, 11,535 feet. A little East of North, exactly above the Sett Sass, a long snow range glowing in the mid-day sun identifies itself with the Antholzer Alps beyond Bruneck, the highest points of which are the Wildgall (10,785 feet), the Schneebige Nock (11,068 feet) and the Hochgall, still, I believe, unascended, and rising to 11,284 feet. Beyond these again, to the N.N.W., Clementi believes that he recognises the Drei Herrn Spitze (11,492 feet) and the Gross Venediger (12,053 feet); these last being full forty-five miles distant as the crow flies.

Turning now from the Northern half of the horizon where all is so clear, it is doubly disappointing to face the mists which still keep pouring up from the South. Parting here and there at times, as if rent suddenly by gusts of wind from the South-West, they show now the tremendous wall of the Cimon della Pala; now the Castelazzo over against the Costonzella pass, and behind the Castelazzo, the Cima d'Asti; and now all the great Primiero peaks in detached glimpses, from the Pala di San Martino to the Sasso di Campo. The Pala di San Lucano, which rises due South of our position, also gleams out now and then, as also does the volcanic cone of Cima di Pape. What might be visible on this side under more favourable circumstances, it is, of course, impossible to say; but I am inclined to think the Southward view, including as it does the Primiero group, would be finer than that from Monte Pavione, which is some 200 feet lower than the Sasso Bianco. As it is, even with one half of the horizon continually obscured, we succeed in identifying over fifty great summits, including all the Dolomite giants. I should be afraid to conjecture how many peaks which could not be verified with certainty must have been in sight.

It was at the time, and is still, a matter of regret to the writer not to have been able to make some kind of panoramic outline, however rough, of the view from the summit. But it would have been useless to make the attempt under such heavy disadvantages, not more than forty-five degrees of horizon being absolutely clear at any time.

As regards the height of the Sasso Bianco, there can, I think, be no doubt that it rather exceeds than falls below 8,000 feet. A traveller more experienced in the use of the aneroid would doubtless be able to determine the matter to within a few feet; but I should, myself, be very diffident of giving a decided measurement. We observed the aneroid closely all the way from Caprile to the summit, and found that it showed a fall equivalent to a rise in elevation of 4,500 English feet. This (without any correction for the mean temperature of the column of air between the upper and lower stations) if added to the height at which Caprile stands above the sea-level, — namely, 3,376 feet—would give an elevation of 8,776 feet. The temperature, however, varied greatly, the heat being intense as we wound round the mountain from East to South, and the change to cold and damp being very sudden when we came into the mists a thousand feet below the summit. These mists never rose to the height of the actual summit during the whole two hours that we remained upon the top. On the contrary, the sun shone uninterruptedly, and the temperature must have stood at from 70° to 75°.

u

Not venturing to deduce results from these imperfect observations, I have submitted my notes to an eminent mountaineer, whose opinion I prefer to give in his own words :—" Assuming the temperature to be respectively 50° and 70°, we should have a correction of 280 feet,* which must be added to your 4,500. The height would then come out $3,376 + 4,500 + 280 = 8,156$ feet ; so that I think you may safely put it at *over* 8,000 feet. In your letter, you spoke of your peak being 400 to 600 feet higher than Monte Migion. Now Trinker gives the latter as 7,838 feet, which would bring the Sasso Bianco up to 8,238 or 8,438 feet ; so that in this way too you get the estimate of over 8,000 feet confirmed." F.F.T.

For the present, then, and until some more competent traveller shall determine this point with accuracy, the height of the Sasso Bianco may be allowed to stand at something over 8,000 feet.

Having spent two hours on the top, and seeing no hope of any change for the better on the Southern side, we reluctantly packed up and came down. By the time we reached Signora Pezzé's pasturages, the Sagro was breaking up in Rocca, and the contadini who lived in the scattered farms and cottages of Monte Pezza were coming up homewards. All asked if we had had a good view ; if we were very tired ; if we had found it difficult ; and how long it had taken us to get to the top.

" Brava ! brava ! " said one old man. " So, Signoras,

---

* The temperature was certainly higher than this at times by five, if not ten degrees, which would bring the number of feet up to 50 or 100 more.

you have been up our mountain? Ebbene! E una bella montagna!... but you are the first forestieri who have cared to find it out."

It was amusing to see how pleased, and even flattered, they all seemed ; as if, being born and bred upon the mountain, they took the expedition as an indirect compliment paid to themselves.

When at length we reached Caprile, it was just half-past five o'clock. We had been gone precisely twelve hours and a half :—that is to say, we had been six hours getting to the top, including stoppages ; two hours on the top ; and four hours and a half, including another stoppage, coming down.

We might, as I have already said, have had a better day. We might, as I fully believe (there being an almost continuous line of valleys, and no mountain range of any importance between), have seen straight down to Venice and the Adriatic on the South ; to the lake of Garda on the South-West ; and perhaps, if the Marmolata is not in the way, to the Ortler Spitze on the West. In any case, the view to the North and North-West was extremely fine ; and the near view over the whole surrounding group of Dolomites (which is of more importance than any distant view of peaks which are continually seen from other heights) is of the greatest interest. I doubt, indeed, if there be any other point from which all the giants of the district can be seen at once, and to so much advantage.

FORNO DI ZOLDO AND ZOPPÉ.

ON THE ROAD AGAIN—NEAR VIEW OF THE CIVETTA—ADVENTURE
WITH A SNAKE—MONTE FERNAZZA—MONTE COLDAI—THE MARMO-
LATA FROM THE PASS OF ALLEGHE—UNEXPECTED VIEW OF THE
PELMO—THE MOUNTAINS OF VAL DI ZOLDO—THE BACK OF THE
CIVETTA—THE VALLEY OF ZOLDO—THE HORRORS OF CERCENA'S INN
—THE SCULPTOR OF BRAGAREZZA—ZOPPÉ; ITS PAROCO, AND ITS
TITIAN—LUNCHEON IN A TYROLEAN COUNTRY-HOUSE—BRUSETOLON
AND HIS WORKS—SPECIMEN OF A NATIVE—VALLEY AND PASS OF
PALLAFAVERA—IN THE SHADE OF THE PELMO—PESCUL—SELVA AND
THE ABORIGINES—CAPRILE AGAIN.

CHAPTER XIII.

FORNO DI ZOLDO AND ZOPPÉ.

THERE remained yet another important excursion to be taken from Caprile, before we could finally break up our camp and depart. We must go over the Pass of Alleghe; visit the Val di Zoldo; * make a pilgrimage to a certain village called *Zoppé*, where a Titian was to be seen, and come home by way of the Val Fiorentino. Now the main attractions of this expedition did not appear upon the surface. We had been over a good many passes already, and through a good many valleys, and had been plentifully pelted with Titians of all degrees of genuineness; but what we really wanted was to see the back of the Civetta, and to get a near view of the Pelmo. As both of these ends would be answered by following the route thus laid down, and as the expedition was guaranteed not to exceed three days, we once more packed our black bags, stocked the luncheon-basket, rose at daybreak one fine morning, and departed. This time, young Cesare Pezzé, the ex-Garibaldian, having a married sister at Piève di Zoldo

---

* Travellers starting from Longarone can now visit the Val di Zoldo and Forno di Zoldo by diligence. (*Note to Second Edition.*)

whom he wished to see, volunteered to walk with us—
a soldierly, upright, picturesque fellow, with his coat
flung loosely across one shoulder, a yellow silk handkerchief tied cornerwise round his throat, a bunch of
carnations in his hat, and an alpenstock in his hand.

This time, as last time, our way lies at first beside
the lake; but strikes away presently behind the village
of Alleghe, and up a delicious little valley thick with
walnuts and limes, and threaded by a bright torrent
that fills many a moss-grown water-trough and turns
many an old brown wheel. The path, rising and winding continually, passes farm-lands and farm-houses;
barns, orchards, gardens; green slopes striped with
rows of yellow flax laid down to bleach in the sun; and
terraces after terraces of wheat, barley, flax, hemp,
potatoes, and glossy-leafed, tassel-blossomed Indian
corn.

And as the path rises, so also rises the Civetta, its
lower precipices detaching themselves in grand proportions from the main mass, while every riven pinnacle,
spire, obelisk and needle-point, stands out sharply
against the deep blue sky. Thus the mountain grows
in grandeur with every upward foot of the way. White
patches that looked like snow-drifts from the valley,
now show as glaciers coated with snow, through which
the blue ice glitters; and by-and-by, as we draw still
nearer, another of those strange circular holes, or
" occhi " as they are here called, stares down at us
from near the top of a small peak, like a hole drilled in
a dagger-blade. So, with exquisite glimpses over the
bluish-green lake, we emerge at length from the gorge,

and climb a steep, stony lane with never a tree on either side to screen off the burning sun.

Suddenly a long steel-blue snake specked with white, darts out from under the very feet of white Nessol! Clementi utters a wild war-whoop—L. a scream—the mule a snort of terror! Giuseppe and young Pezzé leap forward with their sticks, and in a second the poor reptile (which is as thick as one's wrist and about four feet in length, but I believe quite harmless) lies dead by the wayside.

The stony path now leads out upon a wild and desolate mule-track skirting the grim flanks of Monte Fernazza—a gruesome mountain whose low black precipices have crashed down before now in many a berg-fall, covering the barren slopes with shattered débris and huge purply blocks all blistered over with poisonous-looking lichens. Winding now round the head of the glen by which we have come up from Alleghe, we arrive at last upon a grassy plateau at the foot of an overhanging cliff which, though locally called the Monte Coldai, is in truth the huge north-eastern shoulder of the Civetta. Above here, in a hollow among the rocks, nestles a small tarn called the Lago Coldai, said to command a fine view, but we had not time to go so far out of our way.

Beyond Monte Coldai, the way lies up a fine rock-strewn gorge, just like the gorge of the Avisio where it leads up to the Fedaja Alp. Gradually we lose sight of the long, fretted façade of the Civetta, which retires behind the Coldai rocks, and, looking back, find that the lake has sunk quite out of sight. The Sasso Bianco, which till now had been standing out against the sky, has all at once dropped below the horizon, and is immeasurably overtopped by the towering altitudes of the Marmolata. The Boé, the Cima di Pape, the Monte Vernale, the Sasso di Val Fredda, and many another now-familiar peak, have also risen into view. But it is the Marmolata that claims all one's attention, and seems to fill the scene. Presently, an obstinate cloud that has been clinging to the highest point of the summit clears off little by little, and leaves the whole noble mass distinctly relieved against the western sky.

"Guardate!" says young Pezzé, seeing a sketch in preparation. " La Marmolata has thrown her veil aside to have her portrait taken."

It is a grand view of the mountain, even though its snows and glaciers are all out of sight. From here, as from the Sasso Bianco, one sees its true form and its actual summit; while of the one no idea can be formed, and of the other no vestige is visible, from either the Tre Sassi or the Fedaja. Clementi can even identify the tiny top-most patch of snow on which F. F. T. placed his barometer when he reached the summit.

And now a grassy Col, about a quarter of an hour ahead, is pointed out as the summit of the pass. There we shall see the mountains of Val di Zoldo, and take our midday rest in whatever shady spot we can find. There too, as young Pezzé pleasantly prophecies, we shall be within reach of a châlet where milk, and even cream, may be purchased. So we press on eagerly, but, stopping suddenly a little below the top, are amazed to see the Pelmo—snow-ridged, battlemented, stupendous—shoot up all at once, as it seems, from behind the slopes and fir-woods to the left of the pass, as near us as the Civetta! Large masses of vapour are rising and falling round those mighty towers, never leaving them wholly uncovered for an

THE MARMOLATA FROM THE PASS OF ALLEGHE.

instant; but they look all the mightier for that touch of mystery.

And now a few yards higher, and the Marmolata, the Sasso Bianco, the Boé, and all the rest, disappear together, and a lovely grassy plain dotted over with strewn rocks and clumps of firs, and bounded by a line of mountain peaks as wild and fantastic as anything we have yet seen, lies spread out in sunshine before us. This, according to the map, must be the grand chain of which Monte Pramper and Monte Piacedel (both as yet unascended) are the dominating summits.

Up here we encamp for an hour and a-half, Sub Jove; and the mules graze while we take luncheon. Clementi vanishes up the hill-side, and returns by and by with a bowl of cream in each hand, which, beaten up with wine and sugar, and eaten in the midst of such a scene, is at least as delicious as the " dulcet creams " prepared by Eve for the Angel's entertainment. Meanwhile the cow-herd comes down from the châlet to stare at the forestieri, and is so overpaid with half a lire that I begin to fear we must have given him a piece of gold by mistake.

A deep, narrow gorge now leads down from a little below the summit of the pass, to a point whence the Val di *Zoldo*—sunny, cultivated, sparkling with villages and spires—opens out far and wide beneath our feet.

And now, at last, we see the back of the Civetta. Accustomed as one has become to the strangely different aspects under which a Dolomite is capable of presenting itself from opposite points of the compass, here is a metamorphosis which the most erratic imagination could never have foreseen. To say that the Civetta is unrecognisable from the *Z*oldo side is to say nothing; for the mountain is so strangely unlike itself that, although one has, so to say, but just turned the corner of it, the discrepancy in form, in character, and apparently also in extent, is almost past acceptance. Calm, perpendicular, majestic on the side of

Alleghe, here it is wild, tossed, tormented, and irregular. From Alleghe, it appears as a vast, upright, symmetrical screen—here it consists of a long succession of huge, straggling buttresses divided by wild glens, the birthplaces of mists and torrents. If from Caprile the mountain looks, as I have said more than once, like a mighty organ, from here it seems as if each vertical pipe in that organ-front were but the narrow end of rock in which each of these buttresses terminates. Looking at them thus in lateral perspective, I can compare them, wild and savage as they are, to nothing save that vista of exquisitely carved and decorated flying buttresses just below the roof of Milan Cathedral, which is known as the Giardino Botanico.

The Civetta was first ascended by Mr. F. F. Tuckett, who gives the height at about 10,440 feet. The summit, snow-crowned and lonely, is plainly seen from this side, and looks as if it might be reached without serious difficulty.

The Valley of Zoldo is richly cultivated ; the farmhouses are solidly built ; and the whole district wears a face of smiling prosperity. The usual little dusty hamlets with the usual religious fresçoes on the principal house-fronts, the usual little white church, and the usual village fountain, follow one another rather more thickly than in most other valleys. At San Nicolo, where the valley narrows and the rocks close in upon the rushing Maé far below, we enter upon an excellent carriage-road which goes from this point, by an immense détour, to Longarone. At a certain village

called Dont, some way below San Nicolo, we had proposed to pass the night; but being daunted by the dirt
and general disorder of the inn, push on for Forno di
Zoldo where Ball's Guide reports "comfortable quarters
at Cercena's inn."

Here we arrive at the end of another three-quarters
of an hour, and alight at the door of a very large, very
old, and very dirty-looking house up a small steep
street in the heart of the village. Passing through a
gloomy stone kitchen where some fifteen or twenty
harvesters are eating polenta out of wooden platters,
we are shown up a dark staircase and into a large
room, the floor of which is encrusted with the filth of
centuries. The sofa, the chairs, the window-curtains
look as if dropping to pieces with age and only held
together by cobwebs. The windows open on a steep
side-lane where all the children in the place presently
congregate, for no other purpose than to flatten their
noses against the panes and stare at us, till candles are
brought, and curtains can be drawn to exclude them.
As for the landing, which in most Tyrolean inns is the
cleanest and smartest place in the house, it is the
dreariest wilderness of old furniture, old presses, old
saddles and harness, sacks, undressed skins, and dusty
lumber of all kinds, that was ever seen or heard of outside the land of the Don.

Yet the Cercenas themselves are well-mannered
superior people, and their forefathers have owned
estates in Val di Zoldo for over five hundred years.
The daughter-in-law of the house, a pretty, refined-
looking young woman, waits upon us, and is made

quite wretched by our few and modest requirements. We are not, I think, unreasonable travellers ; but we have been riding and walking for nearly twelve hours, and wish, not unnaturally, for water, towels, food, and coffee. For all these things we have to wait interminably. That we should require a table-cloth is a serious affliction, and that we cannot sup, like the haymakers, off polenta, is almost more than young Signora Cercena knows how to bear. A few small lumps of smoke-blackened meat, a dish of unwashed salad, and some greasy fritters are at length brought ; and this young lady, while professing, I imagine, to wait at table, walks over quite coolly to a looking-glass at the farther end of the room, and there deliberately tries on L.'s hat and all my rings and bracelets.

It is a dreadful supper, and is followed by a dreadful night—hot, and close, and wakeful, and enlivened in a way that has associated Forno di Zoldo, for ever in my mind with that Arab proverb which describes Malaga as a city "where the fleas are always dancing to the tunes played by the mosquitoes."

The mules are brought round early next morning, for we have a long day before us. Zoppé, distant rather more than three hours from Forno di Zoldo, has to be visited in the morning ; and at two P.M., on our way back, we have promised, in compliance with Signora Pezzé's particular request, to call upon her married daughter who lives at Pière di Zoldo, about a quarter of an hour above Forno. While we are at breakfast, it being a little after five A.M., the church-bells ring out a merry peal. Concluding that it is either a

Saint's-Day or a wedding, I enquire what the joyful occasion may be, and learn, not without surprise, that an old and highly respected inhabitant has just given up the ghost.

The Val di Zoppé, sometimes called the Val di Rutorto, branches away from the Val di Zoldo at an acute angle from a point a little below Forno, and runs off northward towards the Pelmo. Our way thither lies at first through a chain of villages—Campo, Pieve, Dozza, Prà, and Bragarezza. Passing Pieve, we are met by Cesare Pezzé who is to take us to the studio of a certain self-taught wood-sculptor named Valentino Gamba. He lives at Bragarezza—a miserable tumbledown hamlet on a steep hill-side a mile or two farther on, where we first catch sight of him sitting in a desponding attitude on the doorstep of a small cottage.

Being addressed by young Pezzé and invited to show his studio, he jumps up in red confusion, and leads the way into a little back room where stands an enormous oval frame of carved pine-wood destined for the Vienna Exhibition of the present year (1873). It is an unwieldy, overdone thing, loaded with Arabesques, fruits, flowers, musical instruments, Cupids, and the like ; too big ; too heavy ; fit neither for a mirror nor a picture ; but quite wonderful as an effort of untaught genius. An ideal bust of Italia, also in wood, is full of sweet and subtle expression, and pleases me better than the frame.

What possesses me, that I should enquire the price of that bust ? It is life-size, and weighs—heaven only knows how much it weighs, but certainly as much as all our scanty baggage put together! I have no sooner asked the unlucky question than, seeing the flash of hope in the poor fellow's face, I reproach myself for having done so. He asks only two hundred lire for it —less than eight pounds—but I could no more be burthened with it on such a journey than with the church steeple. So I ask for his card, and, promising to bid my English friends look out for his frame next summer in Vienna, take my leave with the awkward consciousness of having said more than I intended.

From Bragarezza, the way lies between forest-clad hills up a constantly rising valley. The farther we go the steeper and rougher the path becomes; the more desolate the valley; the more noisy the torrent. Then at last we have to dismount and let the mules scramble on alone. Now the Pelmo, as yesterday, comes suddenly into sight; its huge, tawny, snow-ridged * battlements rising close behind a near hill-side—so close that it seems towering above our heads. And presently—for we are only just in time to see it clearly for a few minutes—a great white cloud sails slowly up from somewhere behind, wrapping the mountain round as with a mantle, so that we only catch flitting, fragmentary glimpses of it now and then, through openings in the mist.

Finally Zoppé, a tiny brown village and white church perched high on a green mountain side, looks down

---

* These snow-ridges, which I have likened elsewhere to the steps of a gigantic throne, are chiefly remarkable on the sides facing Zoppé and the Val d'Ampezzo. From the Val Fiorentino they are much less observable.

upon us from the top of a steep path full 400 feet above the valley.

That little white church contains the Titian which is the glory of all this country-side. A long pull up the hill in broiling sunshine brings us at last to the houses and the church. The door stands open, and, followed by all the men out of a neighbouring wood-yard, we pass into the cool shade within. There, over the high altar, hangs the Titian, uncurtained, dusty, dulled by the taper-smoke of centuries of masses. It is a small picture measuring about four feet by three, and represents the Virgin and Child enthroned, supported by San Marco and San Girolamo, with Santa Anna sitting on the steps of the throne. It is, on the whole, a perplexing picture. The Madonna and child, painted in the dry, hard style of the early German school, look as if they could not possibly have come from Titian's brush; the San Girolamo and Santa Anna scarcely rise above mediocrity; but the head and hands of San Marco are really fine, and go far to redeem the rest of the picture. The colour, too, is rich and solid throughout.

This altar piece, painted, it is said, by order of one of the Palatini in 1526, is classed by Mr. Gilbert among the "very few indubitable Titians" yet preserved among the painter's native mountains; but notwithstanding its reputation, I find it difficult to believe that the great master painted much more than the head and hands of San Marco.

The Paroco, hearing that there were strangers in the church, came presently to do the honours of his

Titian. He was a fat, rosy, pleasant little priest, redolent of garlic, and attired in light-blue shorts, a light-blue waistcoat, grey worsted stockings, and a long black clerical coat, worn bottle-green with age. He chattered away quite volubly, telling how Titian had once upon a time come up to Zoppé for villeggiatura in time of plague; and how he had then and there painted the picture by order of the aforesaid noble, who desired to place it in the church as a thank-offering; also how it had hung there venerated and undisturbed for centuries, till the French came this way in the time of the First Napoleon, and threatened to rob the Commune of their treasure, whereupon the men of Zoppé made a wooden cylinder, and rolled the picture on it, and buried it in a box at the foot of a certain tree up in the forest.

"And look!" said the Paroco, "you may see the marks of the cylinder upon the canvas to this day. And we have the cylinder still, Signora—we have the cylinder still!"

I said something, I no longer remember what, to the effect that a genuine Titian was worth taking care of, and that the Commune could not value it too highly.

"Value it!" he repeated, bristling up rather unnecessarily. "Value it, Signora! Of course we value it. Many governments have offered to buy it. We could sell it for three thousand gold ducats to-morrow, if we chose. Ebbene! we are only six hundred souls up here in the Paese. Our men are poor—all poor—contadini in summer, legnatori in winter; but no price will purchase our Titian!"

X 2

We afterwards learned that this public-spirited little Paroco had been a mighty chamois-hunter in his youth, and one of the first to scale the fastnesses of the Pelmo.

Now we leave *Zoppé* on its hill-side and come down again into the valley, catching by the way some wonderful glimpses of strange peaks peeping out through mist and cloud in the direction of Monte Sfornioi and the Premaggiore range. And now, after a brief halt in the shade of a clump of trees beside a spring, we go on again, descending all the way, till we find ourselves back at Piève di Zoldo and alighting at the gate of a large white house, where we are welcomed by young Pezzé's sister, Signora Pellegrini. Now Signora Pellegrini has married a man both wealthy and well descended, and lives in a large, plentiful, patriarchal way, much as our English gentry lived in the time of the Tudors. She carries her keys at her girdle, and herself superintends her dairy, her cows, her pigs, her poultry, and her kitchen. Being ushered up a spacious staircase, and across a landing hung with family portraits of Pellegrinis who were once upon a time Bishops, Priors, Captains, and powdered Seigneurs in ruffles and laced coats, we are shown into a reception room where a table is laid for luncheon.

The master of the house is unavoidably absent, being gone to a cattle-fair at Longarone ; but Cesare Pezzé takes his place at table, where everything is fresh, abundant, home-made, and delicious.

After luncheon, we go to see the church—a large structure with a fine Gothic nave, containing two or three curious early Italian pictures, and an important carved altar-piece by Andrea Brusetolon, the Grinling Gibbons of South Tyrol, born in this valley of *Zoldo* in the year 1662. It is a quaint, strange subject, admirably executed, but not pleasant to look upon. They call it the Altare degli Animi, or Altar of the Souls. Two figures intended to represent Human Suffering and Human Sorrow, each attended by a warning skeleton, support the entablature on each side. Two angels and a Pietà crown it on the top. The execution is excellent, but the impression produced by the work is infinitely painful.

That evening we wander about the fields and lanes beyond the village, and the writer sketches some wild peaks (called by some the Monte Serrata, and by others the Monte Rochetta) which are seen from every point of view about the place. There is, of course, the customary difficulty of keeping intruders at bay. One old woman in wooden clogs, having looked on for a long time from her cottage-door, comes hobbling out, and surveys the sketch with a ludicrous expression of bewilderment.

"Why do you do that?" she asks, pointing with one skinny finger, and peering up sidewise into my face like a raven.

I answer that it is in order to remember the mountain when I shall be far away.

"And will *that* make you remember it?" says she, incredulously.

To this I reply that it will not only answer that purpose, but even serve to make it known to many of my friends who have never been here. This, however, is evidently more than she can believe.

"And where do you come from?" she asks next—after a long pause.

MONTE SERRATA.

"From a country you have no doubt heard of many a time," I reply. "From England."

"From England! Jesu Maria! From England! And where is England? *Is it near Milan?*"

Being told that it is much more distant than Milan and in quite the opposite direction, she is so confounded that she can only shake her head in silence, and hobble back again. When she is half-way across the road, however, she stops short, pauses a moment to consider, and then comes back, armed with one last question.

"Ecco!" she says. "Tell me this — tell me the truth—why do you come here at all? Why do you travel?"

To this I reply, of course, that we travel to see the country.

"To see the country!" she repeats, clasping her withered hands. "Gran' Dio! Have you then no mountains and no trees in England?"

That evening when we are at supper, Giuseppe comes up to say that the young sculptor is below, having brought the bust over from Bragarezza, to know if I will make him an offer for it. Having brought the bust over! I picture him toiling with it along the dusty road — I see him as I saw him this morning, pale, anxious-looking, out-at-elbows; and for the moment I feel as if it were my fate to yield, and buy. Seeing me waver, L. pronounces me a dangerous lunatic, and even Giuseppe ventures respectfully to represent that if the Signora were really to purchase the "testa di legno" we should in future require an extra mule to carry it. So—not daring to see him, lest I should commit the foolish deed—I send down a polite refusal, and hear of the poor fellow no more.

We are off again next morning by half-past five, thankful to see the last of Forno di Zoldo, with its filthy inn, its forges, and its noisy iron-trade. Far down by the torrent-side in the steep hollow below the village, there may be seen long rows of workshops whence the smoke of many fires is always rising. Here the men of Zoldo, who are for the most part blacksmiths, have made nails from time immemorial, sending their goods down on mule-back to Longarone, and getting up in the same way stores of old iron from Ceneda, Conegliano, and even Venice.

We are returning to-day to Caprile by a pass leading from the head of the Val di Zoldo into the head of the Val Fiorentino, winding round the foot of the Pelmo between that mountain and the Monte Crot. For the first four hours of the journey, we are simply retracing our route of the day before yesterday. A little beyond Dont (whence there is an easy and interesting way to Agordo by the Val Duram) the Pelmo rises up, pale, and shadowy, and most " majestical "; while at San Nicolo the Civetta comes into sight again, half-hidden in rolling, silvery mists. Beyond Marezon, about half way between that village and Pecol, the roads divide, and we turn off from the Val di Zoldo up a long, grassy, undulating valley lying between the Pelmo, the Monte Crot, and the back of the Monte Fernazza. This valley, known as the valley of Pallafavera, is the common property of the Communes of Marezon and Pecol, who divide the pastures equally.

Between the Monte Crot—a small but finely-shaped pyramidal mountain—and the Pelmo, which from here looks like a cloudy Tower of Babel, there rises a long green slope leading to the top of the pass. Here, in the shade of a big tree, on a grassy knoll, we call our first halt. The saddles are taken off, and serve for chairs ;

MONTE PELMO.

[F. 349.

a running spring close by among the bushes supplies us with clear water; and Clementi again fetches cream from a milk-farm a little farther on. So, sitting in the open air, under the bluest of blue skies, eating cream with wooden spoons out of a wooden bowl, we take our rest in as purely pastoral a fashion as the heart of even the fair Scudery could have desired.

The journey to-day is a long one, and it will be necessary to let the mules rest again by and by; so we presently go on again, and at about midday reach the top of the pass, which is called by some the Passo di Pallafavera, and by others the Forcella Staulanza. Hence the path winds down among scattered pines and larches to the very base of the Pelmo. At first the great Dolomite shows as only one stupendous tower; then the second tower, till now hidden behind the first, comes gradually into sight; lastly, they divide, showing a dip of blue sky between. Every turn of the path now brings us nearer, so that the huge mass, rising ledge above ledge, steep above steep, seems to hang above our heads and shut out half the sky.

And now, being within two hundred feet of the base of the mountain, we realise, as nearly as it is possible to do so without attempting any part of the ascent, its amazing size, steepness, and difficulty. We are so near that a chamois hunter could hardly creep unseen along one of those narrow ridges three or four thousand feet above; and yet the extent of the whole is so enormous that a woman following a path leading across yonder slope of débris, looks a mere speck against the rock.

The Pelmo blocks the whole end of the Val Fiorentino. The path leading over the low ridge just opposite is the Forcella Forada (6,896 feet) leading direct to San Vito at the foot of the Antelao in the Val d'Ampezzo. It corresponds in position to the Forcella Staulanza over which we have just come from the foot of the Civetta in Val di Zoldo. The height of the Pelmo, so far as has yet been ascertained, appears to be 10,377 feet; that is to say, it is within a few feet the same as that of the Civetta, and scarcely 200 feet below the summit of the Antelao. The mountain has been repeatedly ascended by the daring chamois hunters of Val di Zoldo, who have discovered four separate ways by which to reach the plateau on the top. It has also been ascended by Fuchs, and by the author of the "Guide to the Eastern Alps," who took it from the Borca side, above the Val Najarone. The two best routes, however, are supposed to be the one from Zoppè, and the one from just above San Nicolo in the Val di Zoldo. Mr. Ball describes the Pelmo as "a gigantic fortress of the most massive architecture, defended by huge bastioned outworks whose walls in many places fall in sheer precipices for more than 2000 feet." He furthermore says, "the likeness to masonry is much increased by the fact that in great part the strata lie in nearly horizontal courses, whence it happens that many of the steepest parts of the mountain are traversed by ledges wide enough to give passage to chamois and their pursuers."

From the Forcella Staulanza, the Monte Rochetta of Val d'Ampezzo and the jagged ridge of the Bec di

Mezzodi, are visible above the slopes of the Forcella Forada. The topmost peak of the Civetta also peers out above the fir-woods bordering the eastern face of Monte Crot; and far away, beyond the sunny vista of the Val Fiorentino, the faint blue peak of the Marmolata is seen against the horizon, its snow-slope outlined in frosted silver.

And now, following the course of the infant Fioretino torrent, we begin to leave the Pelmo behind at every step. One by one, the villages of Pescul and Selva, the Col di Santa Lucia, the Monte Frisolet, the Sasso Bianco, come into view. Stopping for a few moments at Pescul, we go into the little church to see a carved tabernacle by Brusetolon—a tiny, toy-like thing, evidently a recollection of the Baldacchino at St. Peter's, supported upon flowery twisted columns, crowned by an elaborate canopy, and enclosing a crucifixion group with figures about three inches in height. Some of the little angels and cherubs clustered outside the canopy are so tenderly conceived and executed as to remind one of the designs of Luca della Robbia. The people of Pescul prize their little shrine, just as the people of Zoppé prize their Titian, and have refused large prices for it.

It is now one o'clock, and we have been upon the road since half-past five A.M. The mules are tired out, and stumble at every step. The Val Fiorentino stretches on interminably, and the village of Selva, where we are to rest for a good two hours, seems never to draw nearer. We get there at last, however, and put up at a small road-side albergo as rough as any we have yet met with, but very clean and airy. Here the tired mules get each a hearty feed of Indian corn; the men, bread and wine; and we, being shown into a whitewashed upstairs room, proceed to light the Etna and brew a dish of Liebig.

The women of the house—and there are four of them —pursue us to this retreat as soon as they have served out the corn and wine below, and stand, wide-eyed and open-mouthed, in a fever of curiosity, watching all we do, as a party of children might watch the movements of a couple of wild beasts in a cage. They examine our hats, our umbrellas, our cloaks, and every individual article that we have laid aside. The Etna stupifies them with amazement. As for L.'s field-glass lying in the window, they eye it askance, taking it evidently for some kind of infernal machine that may be expected to go off suddenly and blow up the whole establishment. They are, in truth, mere savages—rosy, hearty, good-natured; but as ignorant and uncivilised as aboriginal Australians.

The biggest and rosiest of the four—apparently the mistress of the house—emerging presently from the first dumbness of her astonishment, pours forth a volley of questions, repeating my answers with a triumphant air, as if interpreting them to the rest, and cross-examining me as eagerly and unsparingly as an Old Bailey counsel. Where did we come from? From Forno di Zoldo! Sì, sì—she knew that—the men down stairs had told her so much. But *before* Forno di Zoldo. Where did we come from *before* Forno di Zoldo? From Caprile! Che! che! she knows that also. But *before* Caprile? Surely

we came from far away—from *lontana?* Per esempio!
—where were we born! In Inghilterra! Madonna!
In Inghilterra!

Here she throws up her hands, and the other three
do the same.

"But have you come like this all the way from
Inghilterra?"

What she means by "like this," it is impossible to
say. She probably supposes we have ridden the two
Nessols the whole distance by land and sea, with one
small black bag each by way of luggage ; but the easiest
answer is a nod of the head.

"Santo Spirito! And alone?—all alone?"

Again, to save explanations, a nod.

"Eh! poverine! poverine! (poor little things! poor
little things!) Are you sisters?"

A shake of the head this time, instead of a nod.

"Are you married?"

Another negative, whereat her surprise amounts
almost to consternation.

"*Come!* Not married? Neither of you?"

"Neither of us," I reply, laughing.

"Gran' Dio! Alone, and not married! Poverine!
poverine!"

Hereupon they all cry "poverine" in chorus, with an
air of such genuine concern and compassion that we
are almost ashamed of the irrepressible laughter with
which we cannot help receiving their condolences.

Being really tired and in want of rest, I am obliged
at last to dismiss both Coryphæus and Chorus, and when
they are fairly gone, to lock them out. So at last we

eat our Liebig in peace, and, being only two hours from home, with plenty of daylight still at our disposal, the writer succeeds in getting a sketch of the Pelmo as it appears through the open window, down at the far end of the valley.

The rest of the journey lies chiefly along the rising verge of Monte Frisolet, passing over the Col di Santa Lucia. A little beyond Selva, we enter upon Austrian territory, leaving it again on the hill-side above Caprile, and reaching home by way of the old familiar zig-zag a little after six P.M.

CAPRILE TO BOTZEN.

CHOICE OF ROUTES—GOODBYE TO CAPRILE—PIÈVE D'ANDRAZ—THE
UPPER VALLEY OF LIVINALLUNGO—LAST SIGHT OF THE PELMO—
THE SELLA MASSIVE—THE CAMPOLUNGO PASS—CORFARA—A COMING
PAINTER—A POPULATION OF ARTISTS—TICINI AND HIS WORKS AT
CORFARA—A PHENOMENON—THE COLFOSCO PASS—THE GRODNER
THAL—THE CAPITAL OF TOYLAND—THE TRADE OF ST. ULRICH—
THE LADIN TONGUE—RELICS OF ETRURIA—THE PUFLER GORGE—
THE SEISSER ALP—THE LANG KOFEL, THE PLATT KOFEL, AND THE
SCHLERN—THE BATHS OF RATZES—DESCENT INTO THE VALLEY OF
THE EISACK—BOTZEN—THE ROSENGARTEN ONCE MORE—FAREWELL

THE ROSENGARTEN, FROM BOTZEN.

[P. 359.

CHAPTER XIV.

CAPRILE TO BOTZEN.

THE time at length came for leaving Caprile—for leaving Caprile, and the Dolomites, and the pleasant untrodden ways of South Eastern Tyrol, and for drifting back again into the overcrowded highways of Italy and Switzerland.

We were to re-enter the world at Botzen. All roads, perhaps, led to Rome, when the Golden Milestone stood in the centre of the known universe. So, too, all these central Dolomite valleys and passes may be said to lead, somehow or another, to Botzen. We had plenty of routes to choose from. There was the comparatively new char-road between Monte Latemar and the Rosengarten, known as the Caressa pass. There was the way by Livinallungo and the Gader Thal to Bruneck, and the rail, from Bruneck to Botzen. Again, we might follow the long line of the Avisio through the Fassa, Fiemme and Cembra valleys, to Lavis, where the torrent meets the Eisack and the road meets the railway, not far from Trent. Or we might make for the Grödner Thal and the Seisser Alp, and strike the Brenner line at Atzwang, a little above Botzen.

We decided upon the last. It had many advantages over the other routes. It would take us first along the whole valley of Livinallungo; show us the Sella Massive from three sides of its vast circumference; carry us to St. Ulrich, which is to South Tyrol in respect of the wood-carving trade what Interlaken and Brienz are to Switzerland; carry us over the Seisser Alp close under the shadow of the Lang Kofel, the Platt Kofel, and the Schlern; give us an opportunity of visiting the Baths of Ratzes; and finally land us at Botzen in about a week, or even less, from the time of starting.

We parted from friends when we parted from the hospitable Pezzés, and went away promising ourselves and them soon to return again to Caprile. The morning at five A.M. was cool and bright; but we had already been waiting some days for more favourable weather, and the sky was still unsettled. The church-bells were ringing as we rode out of the village, and the usual procession of remonstrance was winding up towards the church. This time, they were going to pray for dry weather.

"Che! che!" said Clementi, contemptuously, "that is the way they do, Signora! The Paroco watches his barometer; and when the rain is near falling, he calls the people together to pray for it. Perhaps it comes down in the middle of the mass. Then he cries 'Ecco il miracolo!'—and, poor devils! they believe it."

As far as Finazzer's little inn at Andraz, our road lay over ground already traversed. Then we crossed

the torrent, left the valley of Buchenstein opening away
to the right, and, skirting now the rising slopes of the
Col di Lana, continued our course up the main valley
of Livinallungo. At the large village known indifferently as Livinallungo and Piève d'Andraz, we paused
for an hour to feed the mules, and were served with excellent coffee in the cleanest of wooden rooms by the
fattest of cheerful landladies. These people also are
Finazzers, and their opposite neighbours, who likewise
keep an inn, are Finazzers; which is the more perplexing as the one albergo is really comfortable, and
the other of doubtful report. The good one, however,
lies to the Eastward; that is to say, to the right of a
traveller coming up from Caprile. The village, which
is the Capoluogo and post-town of the district, hangs
on the verge of a steep precipice, and stands nearly
1,500 feet higher than Caprile. The view from the
church-terrace is quite magnificent, and not only commands the deep-cut course of the Cordevole from its
source at the head of the valley down as far as Caprile,
but brings in the Civetta, the Marmolata, the Monte
Padon (or Mesola), the Sella Massif, and a host of
inferior peaks.

From Piève d'Andraz as far as Araba—a dismal
looking wooden hamlet at the foot of the slopes below
the south-eastern precipices of the Sella—the valley
rises slowly and steadily. As it rises, it becomes
barren and uninteresting. The jagged peaks of Monte
Padon, emerging gradually from their hood of sullen
clouds, show purply-black against the sky. By and
by, the winding way having brought us, somehow,

Y 2

in a line with the Val Fiorentino and higher than the intervening slopes of Monte Frisolet, we are greeted with an unexpected view of the Pelmo. Shadowy, stately, very distant, it closes the end of an immensely long and glittering vista. We see it for a few moments only, and for the last time. As the path trends inward, it vanishes—as the Civetta and the Marmolata have by this time also vanished. We shall see them no more in the course of the present journey; and who can tell when, if ever, we shall see them again?

And now the huge Sella takes all the horizon—a pile of thickset, tawny towers, like half a dozen stumpy Pelmos clustered together. The mass seems naturally to divide itself into the five blocks respectively entitled the Boé, or Pordoi Spitze, closing the head of the Fassa Thal; the Sella Spitze, looking up the Grödner Thal; the Pissadu Spitze overhanging the Colfosco pass; the Masor Spitze facing Corfara and the Gader Thal; and the Campolungo Spitze, dominating the Campolungo pass, which we are now approaching. As we strike northwards up the bare Col to the right, leaving Araba and the Vale of Livinallungo far below, we have these huge, impending bastions always upon the left.

The trees up here are few and stunted. The Alpine roses are over, and only the bare bushes remain. The golden lilies, the gentians, the rich wild flowers that made most of the other passes beautiful, are all missing; and only a few scant blooms of Edelweiss hide themselves here and there among the moss-grown boulders. The mowers are at work, however, on the slopes, getting in the meagre hay-harvest, and singing at their work. First one voice, then another, takes up the Jodel. It is echoed and flung back from side to side of the valley, now dying away, now breaking out again, sweet, and liquid, and wild as the notes of a bird —of which, no doubt, all these Swiss and Tyrolean melodies were originally imitations.

Now, as we near the top of the Col, new mountains come rising on the northern horizon ;—the Santa Croce, or Heiligen Kreutz, a long mountain terminated towards the west with a couple of twin peaks, like a Cathedral with two short spires ; the dome-shaped Verella Berg ; and the Sass Ungar, or Sassander Kofel, which is in reality an outpost of the Guerdenazza Massif.

Just as we have reached the top of the pass and begun to descend, a long, rumbling peal of distant thunder rolls up from the Livinallungo side, and, looking back, we see the clouds gathering fast at our heels. Down below, in a green, lonely hollow, lies Corfara ; consisting of about a dozen houses and a tiny church. The way is steep, and soft, and slippery—the mules can hardly keep their feet—the storm is coming up. So we hurry, and slide, and stumble on as quickly as we can, and arrive presently in the midst of thunder and lightning at the door of Rottenara's albergo.

The little hostelry consists of two houses, an old and a new. The new house is reserved for travellers of the better class, and contains neither public room nor kitchen. The family occupy the old house, cook in it, and there entertain the guides and peasant-travellers. The new house is made of sweet, fresh, bright pine-wood. The upstairs rooms are all wood—floors, walls, and ceilings alike. The ground floor rooms are plastered and whitewashed.

Who would have dreamed of finding Art in such a place? Who would have dreamed that the grave old peasant covered with flour-dust who just now led the mules to the stable, was the father of a young painter of unusual promise? Yet it is so. Franz Rottenara, the son of our host, is an art-student at Vienna. The house is full of his sketches. The first thing one sees on going upstairs is a full-length figure of Hofer on the landing, done on the wall in colours, life-size, admirably drawn, with a banner in his right hand, and his rifle slung to his shoulder. In the largest bedroom, one end of which serves for a dining-room, hang some capital oil studies of still-life, and several clever heads in crayons. And down below, in a sort of lumber room where the wet cloaks are hung to dry, every inch of whitewashed wall is covered with graffiti—heads, arms, hands, caricatures, full-lengths, half-lengths, Frederic the Great, Goethe, Schiller, Mignon, Mephistophiles, Hamlet, the Torso of the Belvedere, the Fighting Gladiator, the Wild Huntsman, and many more than I can remember or enumerate. The pretty little mädchen who serves our dinner is never tired of answering questions about " mein Bruder zu Wien." He painted those two still-life pictures when he was here last summer, and the Hofer fresco four years ago. He was always drawing, from earliest boyhood, and he studied at Munich before he went to Vienna. He is at home now—came home last night to serve his annual

month with the Corfara rifle-corps—and has just gone
over the hill to see friends at some neighbouring
village.

Later in the day, when he returns from " over the
hill," the young artist, at my request, pays us a visit.
He is not yet five-and-twenty, and is as shy as a girl.
We talk a little about art; but as Herr Franz is not
very strong in Italian, and as the writer's German is
limited, our æsthetic conversation is necessarily somewhat dislocated. I gather enough, however, to see
that he has all the steady industry, the patient am-
bition, and the deep inward enthusiasm of a German
art-student; and I believe that he is destined to make
his mark by and by.

Corfara is, of course, over the Austrian border, and
its people are as thoroughly Austrian as if Campidello
and Caprile were not each within a few hours' journey.
Herr Franz is the only member of his family who
speaks Italian. Neither old Rottenara, nor his
daughter, nor any soul about the village, except the
priest, understands a syllable of any language but their
own.

The great surprise of Corfara, however, is its church.
It is not wonderful, after all, to find a solitary genius
springing up here and there, in even the wildest soil.
Not many miles from Titian's birthplace we found the
Ghedinas. In the valley where Brusetolon was born,
we came upon the young wood-sculptor of Bragarezza.
It is not therefore so surprising that Corfara should
produce its painter. But it is certainly somewhat
startling, when—having strolled out by and by after the

storm has tailed off into a dull drizzle—we peep in at yonder tiny humble-looking church, and find ourselves in the midst of the most lavish decorations. Here, where one would have expected to find only whitewash, are walls covered with intricate mediæval diapering; shrines, altars and triptychs loaded with carved and painted saints, and gorgeous with profuse gilding; stalls, organ-loft, and seats elaborately sculptured; all in the most ornate style of early German Gothic; all apparently new; all blazing with burnished gold and glowing with colour.

The sight of this splendour is so amazing that for the first few minutes one can only wonder in silence; and that wonder is increased when, happening presently to meet the priest, we learn from him that all these adornments are the work of the peasant population of the place—of those very haymakers whom we heard singing this morning in the hay—designed by them, carved by them, painted by them, gilded by them; and the pious free-will offering of their hands. It is a small place, and the inhabitants do not number more than 260 or 270 souls, children included; "but," says the priest, smiling, "they are all artists."

He is a gentlemanly priest, and expresses himself in "very choice Italian." He speaks of Corfara in a smiling, well-bred, deprecating way, as "a lost out-of-the-world spot," and of the church, though one would think he must feel proud of it, as "pretty for so poor a place." When I praise the decorations, he shrugs his shoulders, as implying that better might have been done with larger means.

" One good thing we have, though," he says, " which the Signoras have not seen ; but which I shall be pleased to show, if they do not mind the trouble of returning."

So we turn back—this interview having taken place just outside the churchyard gate—and, re-entering the church, follow him to the back of the altar, where are a pair of painted doors now folded back out of sight, but brought round to the front, he says, in Lent, and closed over the face of the altar. These paintings represent the decollation of Saint Catherine, to whom the church is dedicated ; and, according to the Paroco, were executed in the XIV. or XV. Century by an Italian artist named Ticini, who, as the story goes (and it is always the same story with these village treasures) being detained at Corfara by stress of weather, painted these pictures and presented them to the church, in return for the hospitality of the priest. Beyond this, our Paroco has nothing to tell. He knows no more than I, who this Ticini was ; when or where he lived ; or whence he came. No mention of him occurs in the comprehensive volumes of Crowe and Cavalcaselle on " Painting in North Italy ; " and his works at Corfara obtain not a line of notice in the pages of Ball's Guide, or of Messrs. Gilbert and Churchill's " Dolomite Mountains." I, who never even heard of him before, can only judge from the style of his work that he was a North Italian of the Bellinesque school. The paintings, which are of course on panel, are executed in a brilliant, crystalline, early style, and recall the work of Memling even more than the work of the Friulian painters. The Saint Catherine, slender, round-faced, and fair, is quite of the German type; while the exquisite finish of the costumes, the delicate use of the gilding, and the elaborate treatment of patterns and textures, remind one of Carlo Crivelli.

Four other paintings, also on panel, representing Saint Catherine and other saints, adorn the front of the altar. These works, deeper in tone, but evidently belonging to the same period, are supposed by the priest to be by some other hand. At all events they are all interesting; while the larger paintings at the back are unquestionably of rare beauty and value.

As we leave the church, two little girls come running after the priest, to kiss his hand; an act of homage which he excuses to us in his apologetic, smiling way, saying that it is the custom here, and that the children are " simple, and mean well."

Being now come to where the paths diverge, he wishes us a pleasant journey, lifts his little skull-cap with a courtly air, and turns away to his own home — a cheerful-looking white house with smart blinds and pots of flowers in the windows, and a fat poodle sitting at the gate.

Returning presently to the inn, just as the drizzle thickens and the light begins to fail, we encounter a Phenomenon. It stands in the little yard between the Albergo and the Dependance, discoursing and gesticulating in the midst of a group composed of the Rottenaras, our guides, and a few miscellaneous men and stable-boys. It wears highlows, a battered straw hat, and a brown garment which may be described

either as a long kilt or the briefest of petticoats. Its
hair is sandy; its complexion crimson; its age anything between forty-five and sixty. It carries a knapsack on its back, and an alpenstock in its hand. The
voice is the voice of a man; the face, tanned and
travel-stained as it is, is the face of a woman. She is
gabbling German—apparently describing her day's
tramp across the mountains—and seems highly gratified
by the peals of laughter which occasionally interrupt
her narrative.

"A guide?" she exclaims, replying to an observation
of some by-stander. "Not I! What do I want with
a guide? I have carried my own knapsack and found
my own way through France, through England, through
Italy, through Palestine. I have never taken a guide,
and I have never wanted one. You are all lazy fellows,
and I will have nothing to do with you. Fatigue is
nothing to me—distance is nothing to me—danger is
nothing to me. I have been taken by brigands before
now. What of that? If I had had a guide with me,
would he have fought them? Not a bit of it! He
would have run away. Well, I neither fought nor ran
away. I made friends of my brigands—I painted all
their portraits—I spent a month with them; and we
parted, the best comrades in the world. Ugh! guides,
indeed! All very well for incapables, but not for me.
I am afraid of nothing—neither of the Pope nor the
Devil!"

Somewhat startled by this tremendous peroration, we
go in, and leave her discoursing; and I don't know
that I have ever experienced a more lively sense of

gratification and relief than when I presently learn that this lady is a German. She is no stranger, it seems, at Corfara, but appears every now and then in this mad fashion, sometimes putting up at the Rottenara's, for several weeks together. She paints, she botanises, and I think they said she writes. Giuseppe, who describes her as a Signora "molto brutta e molto allegra," tells next day how she supped that night at the guides' table, and entertained them hugely.

The way from Corfara to St. Ulrich lies along the Gader Thal, through the village of Colfosco, and up a high and lonely valley between the Guerdenazza and Sella Massifs.* There was not a living soul in Colfosco as we rode through—nothing but a ghastly, attenuated Christ against a house-side, nearly as large as life, and splashed horribly, as if with blood, from head to foot. The whole village was out on the hills,—

> " The oldest and youngest
> At work with the strongest,"

getting in the hay.

From above Corfara, and as far as the top of the pass, our path lay close under the tremendous precipices of that part of the Sella known as the Pissadu Spitze. The mountain on this side assumes magnificent pro-

---

* The entire area of the Guerdenazza Massif is estimated at about twenty-two square miles, and its level at something over 9000 feet above the level of the sea. The Sella Massif cannot cover an area of less than fourteen square miles. The principal summit of this latter, *i.e.*, the Boé, or Pordoi Spitz, ascended by Dr. Grohman, is by him given at 10,341 feet. I am not aware that any of the other four summits have been scaled. In superficial extent, the Guerdenazza and Sella Massifs exceed all other Dolomite blocks.

portions, preserving always its characteristic likeness to a Titanic fortress, and showing now and then, through clefts in those giant ramparts, glimpses of a great snowy plateau within, with here and there a blue fold of down-ward-creeping glacier, or a fall of misty cascade. As we mount higher, the last patches of corn and flax give place to a broad, desolate space of boggy turf inter-sected by a network of irregular cattle-tracks, and scattered over with scores of wooden crosses. These mark where travellers have been found dead. They say at Corfara that this Colfosco Col is the most dangerous of all the Dolomite passes, and that the wind in winter rages up here with such fury that it drives the snow and sleet in great clouds which bury and suffocate men and cattle in their progress. There is also no defined path, and the bog is everywhere treacherous.

And now, the summit reached and passed, the Lang Kofel rises on the left above woods and hill tops—a vast, solitary tower with many pinnacles. A sheltered gorge thinly wooded with fir-trees opens before us ; the long-impending rain begins again, hard and fast ; and the path becoming soon too steep for riding, we have to dismount and walk in a pelting storm down a steep mountain-side to Santa Maria Gardena, which is the first hamlet at the head of the Grödner Thal. Here we put up at a tiny osteria till the sky clears again, and then push on for St. Ulrich.

Our way now lies along the Grödner Thal, green and wooded and sparkling with villages. The Sella is gradually left behind. The Lang Kofel becomes more lofty and imposing. The Platt Kofel, like a half-dome, rises into view. The wooded slopes of the Seisser Alp close in the valley on the left; and the Schlern, seen for the first time through a vista of ravine, shows like a steep, black wall of rock, flecked here and there with snow.

Every last trace of Italy has now vanished. The landscape, the houses, the people, the names and signs above the doors, are all German. The peasants we meet on the road are square-set, fair, blue-eyed, and boorish. The men carry wooden krazen on their backs, as in Switzerland. Unmistakeable signs and tokens now begin to tell of the approach to St. Ulrich. The wayside crucifixes are larger, better carved, better painted, and some are picked out with gold. By and by we pass a cottage outside the door of which stands a crate piled high with little wooden horses. In the doorway of another house, a workman is polishing an elaborately carved chair. And presently we pass a cart full of nothing but—dolls' legs; every leg painted with a smart white stocking and an emerald-green slipper!

And now the capital of Toy-land comes in sight—an extensive, substantial-looking hamlet scattered far and wide along the slopes on the right bank of the torrent. The houses are real German Tyrolean homesteads, spacious, many-windowed, with broad eaves, and bright green shutters, and front gardens full of flowers. There are two churches—a little old lower church, and a large, smart upper church, with a bulbous belfry tower painted red. And there are at least half a dozen inns, all of which look clean and promising. The whole place, in short, has a bright, prosperous, commercial air about

it, like a Swiss manufacturing town.   Here, at the
Gasthaus of the White Horse, we are cordially received
by a group of smiling girls, all sisters, who show us
into excellent rooms, give us roast-beef and prunes for
supper, and entertain us with part-songs and zitter-
playing in the evening.

That night there came another thunderstorm followed
by three days of bad weather, during which we had more
time than enough for enquiring into the curious trade of
the place, and seeing the people at their work.

For here, as I have said, is the capital of Toy-land.
We had never even heard of St. Ulrich till a few weeks
ago, and then but vaguely, as a village where wooden
toys and wayside Christs were made ; and now we find
that we have, so to say, been on intimate terms with
the place from earliest infancy.   That remarkable
animal on a little wheeled platform which we fondly
took to represent a horse—black, with an eruption of
scarlet discs upon his body, and a mane and tail derived
from snippings of ancient fur-tippet—he is of the purest
Grödner Thal breed.   Those wooden-jointed dolls of all
sizes, from babies half an inch in length to mothers of
families two feet high, whose complexions always came
off when we washed their faces—they are the Aborigines
of the soil.   Those delightful little organs with red pipes
and spiky barrels, turned by the hardest-working doll
we ever knew ; those boxes of landscape scenery whose
frizzly cone-shaped trees and red-roofed houses stood
for faithful representations of " Tempe and the vales of
Arcady " ; that Noah's ark (a Tyrolean homestead in a
boat) in which the animals were truer to nature than

their live originals in the Zoological Gardens ; that monkey, so evidently in the transition stage between man and ape, who spends his life toppling over the end of a stick ; those rocking-horses with an arm-chair fore and aft ; that dray with immovable barrels ; those wooden soldiers with supernaturally small waists and triangular noses—all these—all the cheap, familiar, absurd treasures of your earliest childhood and of mine— they all came, Reader, from St. Ulrich ! And they are coming from St. Ulrich to this day—they will keep coming, when you and I are forgotten. For we are mere mortals ; but those wooden warriors and those jointed dolls bear charmed lives, and renew for ever their indestructible youth.

The two largest wholesale warehouses in the village are those of Herr Purger, and of Messrs. Insam and Prinoth. They show their establishments with readiness and civility ; and I do not know when I have seen any sight so odd and so entertaining. At Insam and Prinoth's alone, we were taken through more than thirty large store-rooms, and twelve of these were full of dolls—millions of them, large and small, painted and unpainted, in bins, in cases, on shelves, in parcels ready packed for exportation. In one room especially devoted to Lilliputians an inch and a half in length, they were piled up in a disorderly heap literally from floor to ceiling, and looked as if they had been shot out upon the floor by cartloads. Another room contained only horses ; two others were devoted to carts ; one long corridor was stocked with nothing but wooden platforms to be fitted with horses by and by. Another room contained dolls' heads. The great, dusk attic at the top of the house was entirely fitted up with enormous bins, like a wine-cellar, each bin heaped high with a separate kind of toy, all in plain wood, waiting for the painter. The cellars were stocked with the same goods, painted and ready for sale.

Now, the whole population of the place, men and women alike, being with few exceptions brought up to some branch of the trade, and beginning from the age of six or seven years, the work is always going on, and the dealers are always buying. It is calculated that out of a population which, at the time of the last census, numbered only 3493 souls, there are two thousand carvers—to say nothing of painters and gilders.* Some of these carvers and painters are artists, in the genuine sense of the word; others are mere human machines who make toys, as other human machines make match-boxes and matches. A "smart" doll-maker will turn out twenty dozen small jointed dolls one inch and a half in length, per diem; and of this sized doll alone Messrs. Insam and Prinoth buy 30,000 a week, the whole year round.† The regular system is for the wholesale dealers to buy the goods direct from the carvers; to store them till they are wanted; and only to give them out for painting as the orders come in from London or elsewhere. Thus the carver's work is regular and unfailing; but the painter's,

* Baedeker for 1888 gives the population as still only 3845.—*Note to Second Edition.*

† This means a doll turned out every three minutes in a working day of twelve hours, which seems almost incredible; yet I but faithfully repeat what was told me on the spot.

being dependent on demands from without, is more precarious.

The warehouses of Herr Purger, though amply supplied with dolls and other toys, contain for the most part goods of a more artistic and valuable kind than those dealt in by Messrs. Insam and Prinoth. All the studios in Europe are furnished with lay-figures large and small from Herr Purger's stores, and even with model horses of elaborate construction. Here also, ranged solemnly all the length of dimly lighted passages, stand rows of beautiful Saints, large as life, exquisitely coloured, in robes richly patterned and relieved with gold :—Saint Cecilias with little model organs; knightly Saint Theodores in glittering armour; grave, lovely St. Christophers with infant Christs upon their shoulders; Saint Florians with their buckets : Madonnas crowned with stars; nun-like Mater Dolorosas; the Evangelists with their emblems; Saint Peter with his keys ; and a host of other Saints, Angels, and Martyrs. In other corridors we find the same goodly company reproduced in all degrees of smallness. In other rooms we have Christs of all sizes and for all purposes, coloured and uncoloured; in ivory; in ebony; in wood ; for the bénitier ; for the oratory ; for the church-altar ; for the wayside shrine. Some of these are perfect as works of art, faultlessly modelled, and in many instances only too well painted. One life-size recumbent Figure for a Pietà was rendered with an elaborate truth, not to life, but to death, that was positively startling. I should be afraid to say how many rooms full of smaller Christs we passed through, in going over the upper storeys of Herr Purger's enormous house. They were there, at all events, by hundreds of thousands, of all sizes, of all prices, of all degrees of finish. In the attics we saw bin after bin of crowns of thorns only.

One day was devoted to going from house to house, and seeing the people at their work. As hundreds do precisely the same things, and have been doing them all their lives, with no ideas beyond their own immediate branch, there was an inevitable sameness about this part of the pilgrimage which it would be tedious to reproduce. I will, however, give one or two instances.

In one house we found an old, old woman at work, Magdalena Paldauf by name. She carved cats, dogs, wolves, sheep, goats, and elephants. She has made these six animals her whole life long, and has no idea of how to cut anything else. She makes them in two sizes; and she turns out as nearly as possible a thousand of them every year. She has no model or drawing of any kind to work by; but goes on steadily, unerringly, using gouges of different sizes, and shaping out her cats, dogs, wolves, sheep, goats and elephants with an ease and an amount of truth to nature that would be clever if it were not so utterly mechanical. Magdalena Paldauf learned from her mother how to carve these six animals, and her mother had learned, in like manner, from the grandmother. Magdalena has now taught the art to her own grand-daughter; and so it will go on being transmitted for generations.

In the adjoining house, Alois Senoner, a fine, stalwart, brown man in a blue blouse, carves large Christs for churches. We found him at work upon one of three-quarters life-size. The whole figure, except the arms, was in one solid block, fixed upon a kind of spit between two upright posts, so that he could turn it at his pleasure. It was yet all in the rough, half tree-trunk, half Deity, with a strange, pathetic beauty already dawning out of the undeveloped features. It is a sight to see Herr Senoner at work. He also has no model. His block is not even pointed, as it would be if he cut in marble. He has nothing to guide him, save his consummate knowledge ; but he dashes at his work in a wonderful way, scooping out the wood in long flakes at every rapid stroke, and sending the fragments flying in every direction. But then Alois Senoner is an artist. It takes him ten days to cut a figure of three-quarters life-size, and fifteen to execute one as large as life. For this last, the wood costs fifteen florins, and his price for the complete figure is forty-five florins ; about four pounds ten shillings English.

In another house we found a whole family carving skulls and cross-bones, for fixing at the bases of crucifixes—not a cheerful branch of the profession ; in other houses, families that carved rocking-horses, dolls, and all the toys previously named ; in others, families of painters. The ordinary toys are chiefly painted by women. In one house, we found about a dozen girls painting grey horses with black points. In another house, they painted only red horses with white points. It is a separate branch of the trade to paint the saddles and head-gear. A good hand will paint twelve dozen horses a day, each horse being about one foot in length ; and for these she is paid fifty-five soldi, or about two shillings and threepence English.

I have dwelt at some length on the details of this curious trade, for the reason that, although it is practised in so remote a place and in so traditional a way, it yet supplies a large slice of the world with the products of its industry. The art is said to have been introduced into the valley at the beginning of the last century ; no doubt, on account of the inexhaustible supply of arollas, or *Pinus Cembra*, yielded by the forests of the Grödner Thal, the wood of which is peculiarly adapted for cheap carving, being very white, fine-grained, and firm, yet soft and easy to work.

The people of St. Ulrich have lately restored and decorated their principal church, which is now the handsomest in South Tyrol. The stone carvings and external decorations have been restored by Herr Plase Oventura of Brixen, and the painted windows are by Naicaisser of Innsbruck. The polychrome decorations are by Herr Part of St. Ulrich ; the large wooden statues are by Herr Mochneght, also of St. Ulrich ; and the smaller figures on the altars and pulpit, as well as the wood-sculpture generally, are all by local artists. Colour and gilding have of course been lavishly bestowed on every part of the interior ; but the general effect is rich and harmonious, and not in the least overcharged. Above the high altar hangs an excellent copy of the famous Florentine Madonna of Cimabue.

The dialect of the Grödner Thal, called the Ladin tongue, is supposed to be directly derived from the original Latin at some date contemporary with the period of Roman rule. It differs widely from all existing dialects of the modern Italian, and though in some points closely resembling the Rhæto Romansch of the Grisons, and the Lower Romanese of the Engadine, it is yet, we are told, so distinctly separated from both by "well-marked differences both grammatical and lexicographical," as to indicate "kinship rather than identity of stock." Those, however, who admit with Steub the unity of the Rhætian and Etruscan languages, and who agree with Niebuhr in believing the Rhætians of these Alps to have been the original Etruscan stock, will assign a still remoter origin to this singular fragment of an ancient tongue. It certainly seems more reasonable to suppose that the tide of emigration flowed down originally from the mountains to the plains, rather than that the aboriginal dwellers in the fertile flats of Lombardy should have colonised these comparatively barren Alpine fastnesses. This view, the writer ventures to think, receives strong confirmation from the fact that a large number of sepulchral bronzes, distinctly Etruscan in character, have been discovered at various times within the last twenty-five years in the immediate neighbourhood of St. Ulrich. These objects, collected and intelligently arranged by Herr Purger, may be seen in his show-room. They fill two cabinets, and comprise the usual articles discovered in graves of a very early date, such as bracelets, rings, fibulæ, torques, ear-rings, weapons, &c., &c. Philologists may be interested in knowing that there exists a curious book on the Grödner Thal and its language, with a grammar and vocabulary of the same, by Don Josef Wian, a native of the Fassa Thal, and present Paroco of St. Ulrich.

From St. Ulrich to the Seisser Alp, the way leads up through a wooded ravine known as the Pufler gorge. Weary of waiting longer for the weather, we start at last on a somewhat doubtful morning, and find the paths wet and slippery, and the mountain streams all turbid from the rain of the last three days. Neat homesteads decorated with frescoed Saints and Madonnas, and surrounded like English cottages with gardens full of bee-hives and flowers, are thickly scattered over the lower slopes towards St. Ulrich. These gradually diminish in number as we ascend the gorge, and after the little lonely church and hamlet of San Pietro, cease altogether.

Hence, a long and steep pull of about a couple of hours brings us out at last upon the level of that vast and fertile plateau known as the Seisser Alp—the largest, and certainly the most beautiful, of all these upper Tyrolean pasture-mountains. Scattered over with clumps of dark fir-trees, with little brown châlets, with herds of peaceful cattle, with groups of haymakers, and watched over by a semicircle of solemn, gigantic mountains, it undulates away, slope beyond slope, all greenest grass, all richest wild-flowers, for miles and miles around. Yonder, to the South-West, the great plateau rolls on and on to the very foot of the Schlern, which on this side looms up grandly through flitting clouds of mists. A low ridge of black and shattered rocks, called the Ross-zähne, or Horse-teeth, from its resemblance to a row of broken teeth in a jaw-bone of rock, connects the Schlern with the North end of the Rosengarten range, as well as with the Southern extremity of the Seisser Alp, and with the ridge out of which rise the Platt Kofel and Lang Kofel. But the Rosengarten is quite hidden in the mists that keep flying up with the wind from the side of Botzen.

The Lang Kofel, however, stern and solitary, with a sculptured festoon of glacier suspended above a deep cleft in the midst of its bristling pinnacles ; and the Platt Kofel,* crouching like an enormous toad, with its back towards the Schlern, show constantly, sometimes singly, sometimes both together, sometimes in sunshine, sometimes in shadow, as the vapours roll and part.

A vast panorama which should comprehend the Marmolata and Tofana, and many a famous peak beside, ought to be visible from here ; but all that side is wrapt in clouds to-day, and only the Sella and Guerdenazza Massifs stand free from vapour. Now and then the curtain is lifted for a moment towards the West, revealing brief glimpses of wooded hills and gleaming valleys bounded by far mountain-ranges, blue, tender, and dream-like, as if outlined upon the sunny air.

* Ball gives the Lang Kofel a height of 10,392 feet, and the Platt Kofel, 9,702 feet. The latter he reports as "easily accessible from Seiss, or more conveniently from Santa Christina in the Grödner Thal." The Lang Kofel was ascended for the first time by Dr. Grohmann in 1869 ; partly ascended by Mr. Whitwell in 1870 ; and again ascended to the highest summit on the 11th of July, 1872, by Mr. U. Kelso, accompanied by Santo Siorpaes of Cortina.

But (apart from the view it commands of its three nearest neighbours, the Lang Kofel, Platt Kofel, and Schlern) the great sight of the Seisser Alp is—the Seisser Alp. Imagine an American prairie lifted up bodily upon a plateau from 5,500 to 6,000 feet in height —imagine a waving sea of deep grass taking the broad flood of the summer sunshine and the floating shadows of the clouds—realise how this upper world of pasture feeds from thirteen to fifteen hundred head of horned cattle; contains three hundred herdsmen's huts and four hundred hay-châlets; supports a large summer-population of hay-makers and cow-herds; and measures no less than thirty-six English miles in circumference— and then, after all, I doubt if you will have conceived any kind of mental picture that does justice to the original. The air up here is indescribably pure, invigo-rating, and delicious. Given a good road leading up from Seiss or Castelruth and a fairly good Hotel on the top, the Seisser Alp, as a mountain resort, would beat Monte Generoso, Albisbrunn, Seelisburg, and every " Sommerfrisch " on this side of Italy out of the field.

The peasants of these parts preserve vague traditions of a pre-historic lake said once upon a time to have occupied the centre of this Alpine plateau; a legend which gains some colour from the fact that were it not for the gap of the Pufler gorge, down which the drainage flows to the Grödner Thal, there would at this present time be a lake in the depression on the summit.

Having wandered and lingered up here for nearly a couple of hours, we at length begin descending by the course of the Tschippitbach, a torrent flowing down the deep cleft which separates the Seisser Alp from the North West face of the Schlern.* Coming presently to a cheese-maker's hut a few hundred feet below the edge of the plateau, we call our midday halt. A bench and table are accordingly brought out and set in the shade ; the good woman supplies us with wooden bowls of rich golden coloured cream ; the mules graze ; the guides go indoors and drink a jug of red wine with the herds-man and his sons ; the mists roll away, and the huge aiguilles of the Schlern start out grandly from above the woods behind the châlet—as if on purpose to be sketched.

From this point down to the Bath-House at Ratzes, the way winds ever through fir-forests which exclude alike the near mountains and the distant view. About half-way down, we pass within sight of the ruined shell of Schloss Hauenstein, once the home of Oswald of Wolkenstein, a renowned knight, traveller, and Minne-singer, who was born in the year 1367 ; fought against the Turks at Nicopolis in 1396 ; was present at the storming of Ceuta in 1415 ; encountered innumerable perils by land and sea in the Crimea, in Armenia, Persia, Asia Minor, Italy, Spain, England, Portugal,

---

* The height of the Schlern is only 8,405 feet ; but it stands up in such a grand solitary way, and its precipices are so bare and vertical, that it looks higher than many a more lofty Dolomite. The easiest ascent is from Völs, and the view from the top, though said not to be so complete as that from the Ritterhorn nearer Botzen, is extremely fine, and comprises the Adamello, Ortler, Oetzthal, and Antholzer Alps. The South Eastern horizon, however, and consequently all the Primiero Dolomites, are concealed by the near mass of the Rosengarten. "No mountain in the Alps has acquired so great a reputation among botanists for the richness of its flora, and the number of rare plants it produces, as the Schlern."—Ball's *Eastern Alps*, p. 484.

and the Holy Land ; and died here in the castle of Hauenstein in the year 1445. He was buried in the church of the famous Abbey of Neustift near Brixen, where his tomb may be seen to this day. His love-

THE AIGUILLES OF THE SCHLERN.

songs, hymns, and historical ballads, are published at Innsbruck, collated from the only three ancient MS. copies extant, one of which belongs to the present Count Wolkenstein, one to the Imperial Library at Vienna, and one to the Ferdinandeum at Innsbruck.

A more rough and primitive place than the little Bath-House of Ratzes it would be difficult to conceive. It lies at the foot of those tremendous aiguilles which we saw just now from the herdsman's châlet; but we have come down some 1800 feet since then, and now find ourselves at the doors of a building that can only be described as two large wooden châlets united by a covered gallery. The bath-rooms occupy the ground-floor, and the bedrooms the two upper storeys. A tiny chapel; a small bowling-ground; one large general Speise-Saal, where eating, smoking and card-playing are going on all day long; and a tumble-down Dépend-ance about three hundred yards off for the reception of the humbler class of patients, complete the catalogue of the attractions and resources of Ratzes. What the accommodation in that Dépendance may be like it is impossible to conjecture; for here in the " Establish-ment," a small bedroom measuring ten feet by eight, containing a straw-stuffed bed, a wooden tub, a chair, a table, a looking-glass the size of a small octavo volume, and no scrap of carpet or curtain of any kind, is the best lodging they have to offer.

The mistress of Ratzes—a lively, clear-headed, business-like widow, with nine children—makes up seventy beds in the Bath-House, and could find occu-pants for seventy more if she had more space. Her customers are for the most part small tradesmen and their families from Botzen, and peasant-farmers from the neighbouring villages. Two springs, one impreg-nated with iron and the other with sulphur, supply these visitors with baths and medicine. There is a priest in daily attendance, but no doctor; and the patients appear to choose their springs at hap-hazard. The baths are of the simplest kind—mere pine-wood boxes coffin-shaped, with wooden lids just reaching to the chin of the occupant, and a wooden shelf inside to support the back of his head. These boxes, ranged side by side in rows of eight or ten, fill a succession of gloomy, low-roofed basement chambers, and look exactly like rows of coffins in a series of dismal vaults. This impression is heightened very horribly when the unwary stranger, peeping timidly in, as I did, through a wide open-door, sees a head solemnly peering up from a coffin-lid in a dark corner, and hears a guttural voice saying in sepulchral accents :—" Guten Abend."

One night at Ratzes is enough, and more than enough, to satisfy the most curious traveller. Of its clatter, its tobacco-smoke, its over-crowded discomfort, its rough accommodation, one has in truth no right to complain. The place, such as it is, suits those by whom it is frequented. We who go there neither for sulphur, nor iron, nor to escape from the overpowering heat of Botzen, are, after all, intruders, and must take things as we find them.

We leave Ratzes the next morning at half-past nine, having to be down at Atzwang by two P.M. to catch the train for Botzen. The morning is magnificent ; but we are all sad to-day, for it is our last journey with the two Nessols. The path winds at first among fir-forests, rounding the base of the great Aiguilles, and passing a ghastly cleft of ravine down which a huge limb of the Schlern crashed headlong, only twelve years back, strewing the gorge, the pastures, and all the mountain slope with masses of gigantic débris.

Now, still and always descending, we pass farms, hamlets and churches; pear and cherry orchards; belts of reddening wheat and bearded barley; and come at last to an opening whence there is a famous view. From here we look over three great vistas of valley— Northward up the Kunters Weg as far as Brixen and the Brenner; Southward towards Trient and the Val di Non; North-Westward along the wide path of the upper Etsch in the direction of Meran. At the bottom of a deep trench between tremendous walls of cliff, close down beneath our feet as it seems, flows wide and fast the great tide of the Eisack. The high-road that leads straight to Verona shows like a broad white line on this side of the river; the railway, a narrow black line burrowing here and there through tiny rabbit-holes of tunnel, runs along the other. A whole upper-world of green hills, pasture alps, villages, churches, corn-lands and pine-forests, lies spread out like a map along the plateaux out of which those three valleys are hewn; and beyond this upper world rises yet a higher—all mountain-summits, faint and far-distant.

From this point the path becomes a steep and sudden *zigzag*. It is all down—down—down. Presently we come upon the first vineyard, and hear the shrill cry of the first cicala. And now the rushing sound of the Eisack comes up through the trees; and now we are down in the valley—crossing the covered bridge—dismounting at the station. Here is Atzwang;

here is the railway; here is the hot, dusty, busy, dead-level World of Commonplace again !

At Atzwang we part from Clementi and the mules, Giuseppe going on with us to Botzen. Clementi is very loth to say goodbye; and L. "albeit unused to the melting mood," exchanges quite affecting adieux with fair Nessol. As for dark Nessol, callous to the last, he shakes his ears and trots off quite gaily, evidently aware that he has finally got rid of me, and rejoicing in the knowledge.

And now, arriving at Botzen, we arrive also at the end of our midsummer ramble. For a week we linger on in this quaint old mediæval town—for a week the pinnacles of the Schlern and the grand façade of the Rosengarten yet look down upon us from the heights beyond the Eisack. As long as we can stroll out every evening to the old bridge down behind the Cathedral, and see the sunset crimsoning those mighty precipices, we feel that we have not yet parted from them wholly. They are our last Dolomites; and from that bridge we bid them farewell.

THE END.

BRADBURY, AGNEW, & CO, PRINTERS, WHITEFRIARS.

Milton Keynes UK
Ingram Content Group UK Ltd.
UKHW030616291024
2435UKWH00030B/181

9 781240 914173